World University Library

The World University Library is an international series
of books, each of which has been specially commissioned.
The authors are leading scientists and scholars from all over
the world who, in an age of increasing specialisation, see the
need for a broad, up-to-date presentation of their subject.
The aim is to provide authoritative introductory books for
university students which will be of interest also to the general
reader. The series is published in Britain, France, Germany,
Holland, Italy, Spain, Sweden and the United States.

André Cailleux

Anatomy of the Earth

Translated from the French by J. Moody Stuart

World University Library

Weidenfeld and Nicolson
5 Winsley Street London W1

Contents

Publisher's note

Some aspects of geology are still very
controversial, especially those concerning the
movements of continents and continental drift.
Chapter six attempts to give an impartial outline
of the main arguments for and against these
theories in the light of generally accepted
scientific opinion. The author would like to
record that his own views can be found in *La
Terre et son Histoire* (*Encyclopédie Bordas* 1967).

Introduction

The planet on which we live was thought of by the Ancients originally as a flat disc; subsequently it was seen to be a ball or sphere. It was only during and after the fifteenth century, however, that man really began to explore the earth's surface, while Antarctica, the seventh continent, has been opened up only during the last hundred years. More recently still, bore-holes drilled to a depth of over 7 kilometres have yielded information about the lower sections of the earth's crust. Attempts are now being made to drill through the crust to the mantle beneath. At the same time, the movements of artificial satellites, observed by astronomers and analysed in the minutest detail by electronic computers, are revealing to us the exact shape of the globe.

These results and achievements are of the greatest interest. Thanks to them we know more about the nature of the planet on which we live. But their practical information is no less great – for marine and aerial navigation, for mineral prospecting, which has now been extended to the ocean floor, for the search for oil and uranium, and for the exploitation of terrestrial heat and other means of natural wealth.

In our examination of the earth's anatomy, we begin with its external features and the structure of the layer at the surface, the crust. Like all large bodies of matter, the earth consists of atoms, and we shall look at the distribution of the earth's chemical elements. We shall then assess what we know of the earth's interior, which has so far defied direct investigation but about which we can speculate with the aid of scientific instruments. How was the earth formed? How did the continents take shape? Have they been drifting around on its surface like rafts, or have they always been where they are now? Have the poles changed positions? How are we to explain mountain folds, volcanoes, trenches and moun-

Figure 1. The earth photographed from an altitude of 100 miles. Such photographs bring out well the curvature of the earth. The narrow dark area, near the top of the horizon, is the Gulf of California.

Table 1 The solar system

Planet	Equatorial diameter (Earth = 1)	Density (Water = 1)	Probable composition	Surface gravity (Earth=1)
Mercury	0·37	5·70 ⎫	ferro-nickel	0·41
Venus	0·956	5·12 ⎬	surrounded by	0·88
Earth	1	5·53 ⎭	silicates	1
(Moon)	0·272	3·34 ⎱	silicates	0·166
Mars	0·532	3·95 ⎰	dominant	0·37
(Asteroids)	<0·05			
Jupiter	11·20	1·33	? solid	2·53
Saturn	9·05	0·685	H_2O	1·06
Uranus	3·7	1·6	CO_2	0·92
Neptune	3·5	2·2	?	0·95
Pluto	?	?	?	?

CO_2 = carbon dioxide
H_2O = water

N_2 = nitrogen
NO_2 = nitrogen peroxide

tains under the sea? These are the sort of questions we shall try to answer in the light of present-day research. Our discussions will thus be as much of the invisible as of the visible parts of the earth and we shall attach as much importance to the ocean floor as to the land surface; indeed, there is more than twice as much of the earth covered by sea as there is dry land.

The earth's place in the universe is minute. It is one of the nine planets which revolve about the sun. The sun is a star and a million stars form a galaxy, our galaxy being the Milky Way. At least fifty million million galaxies are known, and perhaps there are five hundred million million in the observable universe. The mass of the earth is 330,000th of the sun's, which is itself five thousand million times smaller than a galaxy's.

Solar heat received (Earth = 1)	Temperature at surface (°C)			Atmosphere	Albedo (reflecting power)
	Min.	Mean	Max.		
6.67	−270		+410	none	0.07
1.91		?		CO_2	0.62
1	− 88	+ 10	+ 58	N_2, O_2 etc.	0.39
1	−170	+ 10	+110	none	0.07
0.43	− 70	?	+ 10	infinitesimal	0.15
0.2		− 70		none	?
0.037		−140	−138	He, H, NH_3, CH_4	0.42
0.011		−160	−153	He, H, CH_4, NH_3	0.45
0.003		− 70	−184	CH_4, H	0.48
0.001		−210	−200	CH_4, H	0.52
0.00065		−220	−220	?	0.07

O_2 = oxygen
CH_4 = methane
He = helium
NH_3 = ammonia
H = hydrogen

Of the planets of the solar system, the earth is among the heaviest, its density being the highest, with an average of 5·52 gm/cc, a little less than the density of iron (see table 1). Its gravitational field is fairly strong. The heat received from the sun is moderate. As a result, the mean temperature at the earth's surface is about 11°C, and so is favourable to life as we know it. The other planets are less fortunate in this respect. Venus, slightly nearer to the sun, has a mean temperature of 92°C beneath its layer of cloud, and perhaps as much as 426°C at the ground, owing to the effective greenhouse formed by the clouds. In the other direction, further from the sun than the Earth, is Mars, with a mean temperature of −30°C. As for the moon, although its mean temperature is close to the 11°C of the earth, its extremes of temperature are very pronounced, ranging from −170°C in the shade to $+100°C$ in sunlight. Living creatures as we know them are unable to withstand such severe conditions. Finally, the earth's atmosphere is unique among the planets, being the only one that consists mainly of nitrogen and oxygen, two of the essential constituents of all living organisms.

1 The earth's external form

The earth's shape

In 1925 an old man in the mountains of Afghanistan said to a young geologist: 'No, the earth is not round; you may prove it merely by walking from Kabul to Isfahan, for on the way you will certainly have to cross mountains'. He was obviously confusing the chance irregularities of the earth's surface with its general form, but we can gain an idea of the latter from the curvature of the sea. The ancient Greeks noticed that this surface was curved as they watched a ship receding, and we can easily repeat the experiment. In the evening, from some moderately high point like the top of a cliff or a building we can see the distant lights of a ship or an island twenty kilometres away. If, however, we go down to the beach, we no longer see them because the curvature of the sea between the lights and ourselves forms an obscuring screen.

The ancient Greeks also noticed that at the moment of an eclipse the edge of the earth's shadow on the moon forms the arc of a circle, whatever the position of the moon in the sky. Only a sphere or some body approaching that shape could give such a shadow, and we can see this if we observe the shadow on a wall caused by a ball or a coin as it is turned between the fingers.

To learn the earth's shape more exactly, two methods are available. One is geometrical, or astronomical, due in principle to the Greek Eratosthenes (276–195 BC). A heavenly body, for example, the sun or a star, is observed from two different places at sea level on a particular meridian (a line upon the earth's surface where the sun is at its greatest height at the same moment), and the angle between the observer's line of vision and the vertical is measured at each place, the difference between the angles or latitudes being obtained by subtraction. Next, the distance between the two places is measured over the earth's surface, hence enabling the length of a degree of latitude to be calculated. If sea level follows the shape of a perfect sphere, all measurements from any region of the globe will give the same answer, since for a sphere equal arcs subtend equal angles. If on the other hand, the shape departs from being spherical, the lengths of a degree of latitude will vary; this turns out to be the case and has been known since the eighteenth century.

The other method of getting to know the detailed shape of the earth is based on mechanics. The bricklayer or the architect uses it without thinking when he handles a plumb-line or spirit level, which indicate the direction of the force of gravity, or the vertical, at each place. The same gravitational force affects the movement of the moon and, in an even more sensitive way, that of artificial satellites. From these, when all corrections have been made, we have the second method of learning the shape of mean sea level, or the geoid, as it is called. The results of the two methods agree: the geoid is not an exact sphere but rather an ellipsoid, flattened at the two poles and expanded at the equator such that the polar radius is slightly less than the equatorial radius. The absolute difference is of the order of 21,400 metres, the relative difference 1/298. The dimensions of the ellipsoid determined before and after the launching of artificial satellites are presented in table 2.

It has even been demonstrated from satellites that the equatorial radii are not all equivalent. The value is 6,378,153 metres in America and 6,378,248 metres in the Indian Ocean, but the difference (95 metres) is 225 times less than that between the polar and equatorial radii (21,382 metres) and this latter difference, that is the polar flattening, remains the dominant trend.

The explanation of the flattening is simple. Gravity is the resultant of two forces, the attraction due to the mass of the earth and the centrifugal force of the earth rotating on its own axis. Centrifugal force is zero on the axis, at the poles, and increases as one moves away from these two positions; the force is a maximum at the equator, which is why this region bulges. We can also see that the earth has reacted to these forces as a deformable body. If it were homogeneous, that is to say, if all parts had the same properties, the relative flattening would be $1/231\cdot7$. Since the flattening is actually $1/298$, the earth is not homogeneous and its interior is more dense than the crust. We shall return to this later.

The mass of the earth, determined astronomically, is 5,876 million million million tons, from which its average density is calculated to be $5\cdot527$ gm/cc.

General distribution of land and sea

One only needs to look at a model of the globe to see that the most striking feature is the division into continents and oceans. The continents are six or seven in number, depending whether one counts Asia and Europe together or separately. In reality Europe is only a slender extension of Asia welded at present along the Ural Mountains. It has not always been so, however, and even in our day Europe is distinct from Asia in many respects. After this, in decreasing order of magnitude come Africa, North America, South America, Antarctica, and Australia. Tens of millions of years ago, Africa was separate from Asia, and North America from South America. We know this from findings of ancient marine shells in the isthmuses of Suez and Panama. We are thus justified in considering all the continents as quite distinct units.

Table 2   The shape of the earth

(m = metres)

	International ellipsoid (1909)	Astrogeodetic ellipsoid (1960)
Equatorial radius	6,378,388 m.	6,378,160 m.
Polar radius	6,356,912 m.	6,356,778 m.
Difference	21,476 m.	21,382 m.
Flattening	$\dfrac{1}{297 \cdot 0}$	$\dfrac{1}{298 \cdot 3}$

The oceans are four in number: the Pacific, the Atlantic, the Indian and the Arctic, all of which are in broad contact with one another. The continents on the contrary constitute four isolated entities: the Antarctic, Australia, the Americas, attached only by the isthmus of Panama, and the Old World, consisting of Europe and Asia, joined at present to Africa by the isthmus of Suez.

The submarine depths around the continents are sufficiently well known for one to see what would happen if the level of the sea were lowered, for instance, by 200 metres, or 2,000 metres. In both cases the general outline would remain much the same. This outline is a fundamental feature of the earth's surface, and not simply due to the chance distribution of water. It is deeply scored in the crust of the earth.

There are relatively minor differences between land and sea. Lowering the sea by 200 metres removes irregularities of the coastline and many nearby islands are united to the continents. These well deserve their name of *continental islands* and examples are Great Britain, Newfoundland, Ceylon, Sumatra, Java, Borneo and the Falkland Islands. Tasmania and New Guinea form a block with Australia. North America joins on to Asia in the region of the Bering Straits. Apart from these barriers, the vast areas of the oceans are linked in a single expanse.

With a lowering of 2,000 metres a bridge joins North America to Europe via Iceland and the Faroes, isolating the Arctic Ocean, and Australia is united through Java to Asia. Apart from this the changes are minor. Greenland joins neighbouring America, Spitsbergen joins Europe, and the three great Oceans, Pacific, Atlantic and Indian, retain their joint outline and their broad intercommunication via the Southern seas. Madagascar and New Zealand are still isolated, as are also the small islands scattered throughout the oceans such as the Azores, Hawaii and Tahiti, which may thus be called *oceanic* islands.

Since the contrast between continents and oceans is the principal physiographic feature of the earth it is worthwhile expressing it more precisely in numbers. At sea level the oceans occupy the greater area, 71 per cent against only 29 per cent for land (continents and islands). This ratio however changes as one moves from the ice of the North Pole to warmer lands and on to the South Pole.

The continents take up 39·4 per cent of the surface in the northern hemisphere but less than half this area in the southern hemisphere (18·6 per cent). Moreover it may be seen that a hemisphere centred on the estuary of the river Loire in France is almost half covered by land (47 per cent) and may be called the *continental hemisphere* whereas the complementary half of the globe, dominated by the Pacific

Ocean, bears only 11 per cent land and is thus the *oceanic hemisphere*.

The six or seven continents and four oceans are not scattered randomly over the earth's surface, but according to certain rules, the first of which was formulated by Bacon in 1645 in his work *Novum Organum* when he said that the continents taper to a point. The actual distribution of the continents has been well summarised by J. M. Chevallier as follows:

1 There is a contrast or anti-symmetry with respect to the equator: as we have just seen, land dominates the northern hemisphere, especially between the 60th and 70th parallels, and sea the southern hemisphere, especially between the 50th and 60th parallels. By a remarkable exception, however, the North Pole is occupied by ocean and the South Pole by a continent, thus preserving the anti-symmetry.

2 Considering the globe again, we may observe a plane through the centre of the earth cutting the surface roughly along the meridian 20° east – 160° west (the Russian scientist Voronov would prefer 15° east – 165° west). On one side this divides Europe and Africa and on the other, the Pacific Ocean. With respect to this meridian, the great continental masses are roughly symmetrical, and this may be clearly seen on a globe.

3 It can also be easily seen from a globe that the plane cutting the meridian at right angles to that above, that is, 110° east – 70° west, also bears some anti-symmetry in that it leaves much of the continental masses on one side: Africa, Europe, nearly all South America, and a large part of Asia. Only Australia and part of North America lie on the other side of it, and even the Pacific, the biggest ocean, is almost entirely contained therein.

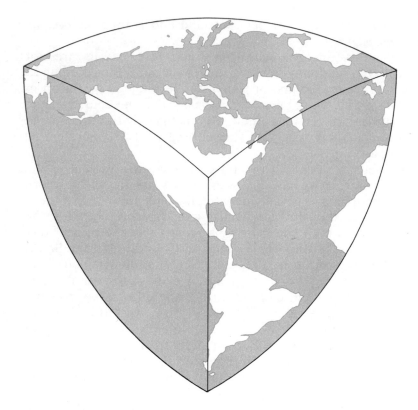

Figure 2. The tetrahedral distribution of continents and oceans. The oceans are envisaged as occupying the faces of a tetrahedron, while the continents occupy the edges.

Figure 3. Overleaf. The far side of the moon, a southern view. Many craters are visible, some as much as a hundred kilometres in diameter. The photograph was taken from a height of 1,400 kilometres above the moon's surface by the U.S. spacecraft Lunar Orbiter–2.

4 Apart from Antarctica at the southern pole of the earth, the other continents fall into three north-south pairs: North America and South America, Asia and Australia, Europe and Africa. These three pairs form boundaries to the three great oceans, the Pacific, Indian and the Atlantic, and they are most distinct in the southern hemisphere where each terminates in a point. This gives a threefold distribution like that of a tetrahedron (figure 2), with the oceans occupying the faces and the continents forming the edges. It is not a truly symmetrical distribution because the angles between adjacent continent pairs are clearly not equal.

5 Still looking at a globe, and rotating it, one can see that each of the aforementioned pairs is, as it were, twisted, with the northern continent displaced to the west and the southern to the east. The angle of displacement varies, reaching 20° of longitude for the Europe-Africa pair, 40° for the Americas and 55° for Asia-Australia-Tasmania. Bounded by these three pairs the three oceans have a similar twist, shown well by the Atlantic and the Pacific.

There is a method of calculation known as harmonic analysis which allows one to express these relationships mathematically from maps. The result thus obtained for each latitude fully confirms the equatorial anti-symmetry, and the first, second, third and fourth harmonics support rules 5, 2 and 3 above as well. Rule 4 does not appear since it concerns a distribution rather than a state of symmetry. An analysis along these lines shows that the first, second, third and fourth harmonics are sufficient to give a very satisfactory representation of the continents and oceans without recourse to higher harmonics. Of the contours calculated in this way, the – 2,000 metres (2,000 metres below sea level) fits the general outline

of the continents best, and this is exactly the one already demonstrated to be particularly significant from the maps.

The five preceding rules are of great importance and any theory of the formation of continents and oceans must provide a reason for their existence. There is a sixth rule to which certain geographers attach a similar importance, namely that every point on land has its antipodal point (opposite point) in the sea, and this is so for 90 per cent of the earth. This rule does hold but it may easily be seen to follow directly from the first five rules. To demonstrate this, let us first consider an extreme case on an imaginary globe where land and sea (in the terrestrial proportions of 29:71) are distributed more or less uniformly, somewhat like the stones in a mosaic. The antipode of any point has then a 71 per cent chance of being in the sea. At the opposite extreme is the case of the globe with all the continents united into a single massive block. Here every point on land must have its antipode in the sea, that is, a 100 per cent chance. The actual situation on earth (90 per cent) is intermediate between the two extremes but closer to the second (100 per cent) than to the first (71 per cent).

Height and depth

Having considered the distribution of land and sea in the horizontal plane, it will be enlightening to view this in the third dimension of soaring mountains and plunging submarine depths. If we glance first at tables 3 and 4 it may be seen that the highest peak in the world, Mount Everest, is 8,848 metres high, and the deepest trench, that of the Philippines, is 11,516 metres under water. The greatest total difference in level known is therefore 20,364 metres. This is of

Table 3 Highest and lowest points on the earth's continental surface

Continent	Mountain	Altitude in metres	Depression	Altitude in metres
Asia	Everest	8,848	Dead Sea	−392
Europe	Elbrouz	5,633	Caspian Sea	− 28
Africa	Kilimanjaro	5,895	Qattara	−130
N. America	McKinley	6,193	Death Valley	− 86
S. America	Aconcagua	6,959	Valdes Peninsula	− 40
Antarctica	Vinson	5,140	Sunk Lake	− 5
Australia	Kosciusko	2,230	Lake Eyre	− 12

precisely the same order as the difference between the equatorial and polar radii of the earth (about 21,400 metres). These 20 kilometres however are negligible in comparison with the 6,368 kilometres of the mean radius, and would correspond to irregularities of 0·2 millimetres on the surface of an orange, which would be practically imperceptible.

The maps in existence are now sufficiently good to enable us to calculate the mean height of the continents above sea level to within a few metres (table 5). For the earth as a whole this has a value of 880 metres, including not only the rock, but also the ice which covers it in several places, especially in Antarctica and Greenland. In the past these ice caps have not always been in existence, and if we correct for the effect of ice, Antarctica is reduced from 2,200 metres to about 500 metres mean altitude, North America from 780 to 690 metres, and the earth as a whole from 880 to 725 metres. In considering the structure of the earth we shall use this last value.

Table 5 however presents evidence of a remarkable

Table 4 The oceanic trenches

Trench	Ocean or sea	Depth in metres	Length in kilometres
Philippines	Pacific	11,516	1,200
Marianas	Pacific	11,033	2,000
Tonga	Pacific	10,882	1,250
Kuril-Kamchatka	Pacific	10,542	2,150
Japanese	Pacific	10,374	1,500
Kermadec	Pacific	10,047	1,200
Bonin	Pacific	9,810	500
Puerto Rico	Atlantic	9,200	800
New Hebrides (north)	Pacific	9,165	500
Solomon	Pacific	9,103	300
Yap	Pacific	8,527	400
New Britain	Pacific	8,320	450
South Sandwich	Atlantic	8,263	1,200
Peru-Chile	Pacific	8,055	1,900
Palau	Pacific	8,054	200
Diamantina	Indian	8,047	160
Aleutian	Pacific	7,679	3,300
Ryukyu	Pacific	7,507	700
Cayman	Caribbean	7,491	900
Java	Indian	7,450	2,400

Table 5 The continents and the oceans

	Surface area in square km.	Mean altitude in metres	Volume in cubic km.
Asia + Europe	54,200,000	860	
Asia + Indonesia	44,200,000	990	
Europe	10,000,000	290	
Africa	29,800,000	660	
N. America (A)	24,200,000	780	
N. America (D)		690	
S. America	18,000,000	650	
Antarctica (A)	13,100,000	2,200	
Antarctica (D)		500	
Australia + Oceania	9,000,000	330	
Sum total (A)	**148,300,000**	**880**	**130**
Sum total (D)		**725**	**107**

		T	T − D
Pacific Ocean	180,500,000	−3,950	−4,330
Atlantic Ocean	92,200,000	−3,540	−3,990
Indian Ocean	75,000,000	−3,870	−3,990
Arctic Ocean	14,000,000	−1,500	−3,050
Sum total	**361,700,000**	**−3,690**	**−1,340**

Continents + oceans	**510,000,000**	**−2,630**

A   denotes the present altitude ; the altitude of the rock/air or the ice/air boundary.

D   denotes the probable altitude of the rock/air boundary if Greenland and Antarctica were deglaciated and the land rose as a result of isostasy.

T   denotes whole of the ocean.

T − D   denotes the ocean excluding the continental shelf.

relationship, known as the Law of Matschinski: the greater a continent, the lower its mean altitude. This is clearly shown in table 5, and must be remembered in envisaging the equilibrium and stability of continents. The actual depth of rock basement beneath the ice in Antarctica is unfortunately known at too few points to say whether Matschinski's Law holds in this case, and it is only here that there is uncertainty. Future research will no doubt resolve this problem.

Until relatively recently it was much more difficult to measure the depths of oceans than the heights of land, but many nautical charts have now been improved by the use of ultrasonic soundings, and the results are shown in table 5. It may be seen that the three main oceans (Pacific, Atlantic and Indian) are very close in the value of their mean depths (3,540 to 3,950 metres) whereas the Arctic Ocean is not as deep as the others, with a value around 1,500 metres.

Frequency of altitude and depth

It is possible to establish from maps the area of the surface of each continent which is above 5,000 metres, and then that between 5,000 and 4,000 metres, 4,000 and 3,000 metres, and so on, and one can do the same for the oceans. If these areas are then plotted against height and depth, we obtain the *hypsographic curve* of the continent or ocean. The hypsographic curves of all the continents including a deglaciated Antarctica are very similar, and so also are those of the oceans. One may rightly group them all into a single world hypsographic curve (figure 4).

Those mountains, so cherished by alpinists and skiers, with heights above 3,000 metres take up less than one per cent of the continental surface, and in the oceans the great

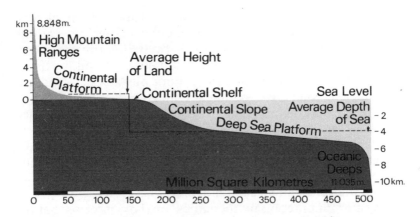

Figure 4. The hypsographic curve shows the relative areas of continents and oceans at various heights and depths above and below sea level. The high mountains and oceanic deeps occupy only a small part of the earth's surface. Most of the land is only a few hundred metres above sea level and most of the sea bed a few thousand metres below.

depths below 7,000 metres occupy even less (0·2 per cent). Two principal levels in particular comprise the greatest area of the earth's surface. One is that of the oceanic depths between 3,000 and 6,000 metres, which includes more than half of the globe. The other level in question comprises the continental plains below 1,000 metres (20 per cent of the earth's surface, and its most heavily populated part) and extends under the shallow edge of the oceans as the *continental shelf*. The edge of this shelf is usually at a depth of 130 metres, or exceptionally, 500 metres around Antarctica and Greenland. Between the two levels just mentioned the intermediate depths of 130 to 3,000 metres take up a mere 5 to 10 per cent of the surface.

The continental shelf is clearly a direct extension of the continents themselves. Its relative area is 6 per cent, which increases the continental part of the earth's crust from 29 per cent to 35 per cent. That part which is below 3,000 metres and is definitely oceanic comprises 57 per cent.

Having combined the areas and the mean altitude and

depths (table 5) it may be seen that the total volume of rock above sea level is about 107 million cubic kilometres, and the volume of water in the oceans is twelve or thirteen times greater, at 1,340 million cubic kilometres. It follows from this that a small variation in the volume of water in the oceans is sufficient to cause a significant transgression or recession of the sea, and there is evidence that this has often happened in the course of the earth's history, as when ice caps were formed on the continents and when they melted again, or when rocks of the crust were uplifted or subsided. The gentle slope of the coastal plains and continental shelves contributes to the wide effect of all such movements.

As a result of recent mathematical analysis it seems that the total hypsographic curve can be resolved into three component normal or bell-shaped curves. The tops of these three, representing the most frequent altitudes, are at values of 500 metres, 200 metres and −4,500 metres respectively. The two extreme maxima at 500 metres and −4,500 metres probably indicate the fundamental difference in origin between the continents and the oceans, while the central one, in between the two chief domains, signifies the double effect of erosion lowering the land and sedimentation filling the sea, which is greatest in both cases near to the shore. On the other hand even the land corresponding to the part of the curve above 500 or 1,000 metres is also affected by erosion.

Continents

If we look again at a globe or an atlas (but choosing a projection which does not deform things too much, that is, preferably not Mercator's), we may see that the continents are very varied in their shapes. Europe is most rich in peninsulas, which occupy twenty-seven per cent of its surface area, and also the most closely integrated, a fact which has facilitated economic and cultural exchanges and helped it to play such a great part in world history. North America is the richest in islands (seventeen per cent of its surface area, due mainly to Greenland) and has a relatively long coastline, stemming from the presence of such large indentations as the Gulf of Mexico and Hudson Bay. If one were to parachute down from an aeroplane at random over a continent, the average distance to the nearest coast would be 780 kilometres in Asia and 340 kilometres in Europe, with the other continents intermediate.

Each of the continents has its great mountain ranges, often surprisingly close to the coasts. This gives a lack of symmetry, which is most marked for Asia where the highest chains are in the south; they are to the east in Australia and to the west in South America. The mountain ranges are often continued as islands or strings of islands or submarine ridges at the coast. The Andes are continued in Tierra del Fuego, the Himalayas and the mountain chains of Burma and Malaya in the Sunda islands, and the Antarctic peninsula in the South Shetland, South Orkney and South Sandwich islands.

Nine-tenths of the great mountain chains are arranged in two main groups; first, along the coasts of the Pacific, which are almost entirely mountainous even in the Antarctic, and secondly, an elongated zone running west to east including the Antilles, and across the Atlantic the Atlas, the Alps, the Carpathians, the Balkans, Asia Minor, the Caucasus, Iran, Pamir, the Himalayas, the Burmese and Malayan ranges and the Sunda islands. At each end, in both Mexico and New Guinea, this zone connects with the circum-Pacific mountain

Table 6 The largest lakes

	Surface area in square km.	Surrounding continent	Maximum depth in metres
Caspian Sea [1]	394,000	Europe & Asia	980
Lake Superior	82,410	N. America	406
Lake Victoria	69,485	Africa	81
Aral Sea	66,460	Asia	68
Lake Huron	59,830	N. America	229
Lake Michigan	58,020	N. America	281
Lake Tanganyika	32,890	Africa	1,435
Great Bear Lake	31,790	N. America	82
Lake Baikal	31,500	Asia	1,741
Lake Nyasa [2]	30,000	Africa	706

[1] this area varies by about 1,000 square kilometres from one year to another.
[2] approximate area.

belt. This zone is of great importance in the earth's structure. It is marked by more or less enclosed seas: the Gulf of Mexico, the Caribbean Sea, the Mediterranean, the Persian Gulf, the Andaman Sea and the Sunda Sea. It earns its title of *Mesogeic zone* from the Greek *meso* meaning middle and *ge* meaning earth; as we have seen, it separates the lands of the north from those of the south, Europe from Africa, Asia from Australia, North America from South America, and it is at this latitude that the continents are obliquely displaced, the northern ones to the west, and the southern to the east.

The fact that the mountains are found along edges of the continents and not in the middle has one further consequence: it favours the existence of inland drainage basins. The Caspian Sea, and the basins of the Sahara and Australia are examples of these. Six per cent of the continental area drains into such basins, and of the remaining ninety-four per cent, forty per cent drains into the Pacific Ocean, thirty-seven per cent into the Atlantic, and seventeen per cent into the Indian Ocean.

Inland seas and lakes (table 6) are extremely useful to man, especially in the fields of public water supplies, fisheries, navigation and leisure activities. Because they occupy less than one per cent of the land surface, and because of their small average depth, the volume of water which they hold is very small, 0·13 million cubic kilometres (including river waters), which is less than one ten-thousandth of the volume of the oceans. However, many of these lakes or inland seas are very interesting in relation to the general structure of the earth because they occupy zones of subsidence, for example the Caspian Sea and Lake Chad, and are often accompanied by fracture of the crust, for example the great lakes of east Africa, and Lake Baikal in Siberia.

Islands

In addition to the continents, a part of the solid land is made up of islands, which are more pleasant the more remote they are, and may have a delightful climate due to the proximity of the sea. Table 7 lists the largest islands. Islands comprise ten and a half million square kilometres, which is only seven per cent of the area of the continents. The largest, Greenland, at 2·2 million square kilometres, is only three and a half times as small as the smallest continent, Australia.

Islands are very diverse in their nature, and we may distinguish the following types:
1 Islands on the continental shelf
2 Islands which are almost continental
3 Large remote islands
4 Island arcs
5 Oceanic islands: volcanic, coral, etc.
Islands on the continental shelf are anchored round the

edge of the continents, and if the sea level were to go down two hundred metres or less, in many cases they would be re-united to their continent, as for example, Great Britain. The geological structure of these islands is often the same as that of the neighbouring continents. In short, they are parts of the neighbouring continent which have become separated only by minor oscillations in sea level and local deformations of the earth's crust. They are never more than 160 kilometres from their continent. In size they are very variable, ranging up to 820,000 square kilometres in the case of New Guinea.

Islands which are almost continental are those which are still very close to a continent but separated by a channel which is definitely deeper than the continental shelf, for example, Ellesmere Island off northern Canada. They possess all the other features of the preceding category. Greenland could perhaps be classed either among these or among the preceding group, depending on whether one defines the continental shelf at a depth of two hundred or five hundred metres, and similarly with Spitsbergen. The latter, however, is situated 400 kilometres from Franz Joseph Land, which is itself that far away from Novaya Zemlya, and this seems to be the maximum distance for such islands.

The group of large remote islands comprises only Mada-gascar (594,000 square kilometres) and New Zealand (152,000 square kilometres for South Island and 115,000 for North Island). Their distance from the nearest continent is great – four hundred to twelve hundred kilometres, and the branch of the sea between the two is deep, two thousand metres or more. We shall see that the rock of these large islands is still continental, and they are really fragments detached from the continents.

Island arcs are strings of islands, or very long, curved

Table 7 The largest islands

		square km.	
1	Greenland	2,175,000	AC
2	New Guinea	820,000	CS
3	Borneo	744,000	CS
4	Madagascar	594,000	LR
5	Baffin Island (Canada)	476,000	CS
6	Sumatra	473,000	CS
7	Honshu (Japan)	228,000	CS
8	Great Britain	230,000	CS
9	Ellesmere Island (Canada)	212,000	AC
10	Victoria (Canada)	212,000	CS
11	Celebes	188,000	IA
12	South Island (New Zealand)	152,000	LR
13	Java	124,000	CS
14	North Island (New Zealand)	115,000	LR
15	Cuba	114,000	IA
16	Newfoundland	111,000	CS
17	Luzon (Philippines)	107,000	CS
18	Iceland	104,000	O
19	Mindanao (Philippines)	95,800	CS
20	Hokkaido (Japan)	89,900	CS
21	Ireland	84,400	CS
22	Novaya Zemlya	79,000	CS
23	Sakhalin (Siberia)	75,400	CS
24	Haiti	74,100	IA
25	Tasmania	67,960	CS
26	Ceylon	65,800	CS
27	Tierra del Fuego	47,900	CS
28	Spitsbergen	39,400	AC
29	Hainan (China)	36,800	CS
30	Formosa (China)	35,800	CS
31	Vancouver (Canada)	32,100	CS
32	Sicily	25,900	CS

IA = island arc   CS = island on the continental shelf
LR = large remote island   AC = an almost continental island
O = oceanic island

archipelagoes situated along the edge of an ocean, with their convex sides towards the open sea. A striking picture of this is shown in figure 6. Most of them are attached by one or both ends to the continent. Certain examples, like the Marianas Arc, seem to float in mid-ocean. These strings of islands encircle the western Pacific in successive arcs which are, running southwards: the Aleutians, the Kuril Islands, Japan, the Ryukyu Islands, the Mariana Islands, the Caroline Islands, Palau, the Philippines. In the Indian Ocean there are fewer arcs, and these are situated in the east, namely, the arc of the Andaman and Nicobar Islands, and the Sunda Islands which have several ranges and swing eastwards towards the Pacific.

The Atlantic has only two island arcs, the Antilles in Central America, and the Scotia Arc, which links South America and the Antarctic peninsula via South Georgia, the South Sandwich Islands, the South Orkneys and the South Shetlands. As well as the islands which are actually visible, between them there are submarine heights and submerged islands along the line of the arcs, such as the Burwood bank in the Scotia Arc which reaches five hundred kilometres in length. Exact measurements have shown that despite appearances, the arcs do not form parts of a circle; the curvature varies, with the radius of one end often twice or three times that of the other – rather like a spiral. The curvature may even change direction, as does the Scotia Arc. The arcs may include islands on the continental shelf like Japan, or be volcanic like the Marianas, or both, like the Sunda and Caroline Islands.

Oceanic islands are situated far out, clearly away from the continental shelf. The largest is Iceland (104,000 square kilometres), followed by Hawaii (10,400 square kilometres).

Figure 5. The volcanic island of San Benedicto, with Boqueron volcano still active. This U.S. Navy photograph was taken in September 1952.

The majority however are much smaller. Some have a sombre and imposing bulk set well above the sea, often forming a cone or a dissected mass. These are volcanoes or groups of volcanoes, which may be either active or extinct, as for example Tahiti and Fiji. Those oceanic islands set in the more pleasant climates of tropical seas are mainly smaller and made of white coral reefs, and often in the shape of a ring, or *atoll*. Recent measurements and borings have shown that deep down these islands rest on a volcanic pedestal. They, too, are ancient volcanoes but long since extinct, and submerged and covered with calcareous structures and corals. The elongated Hawaiian archipelago shows a really remarkable change from east to west: the volcanoes of the eastern end are emergent and active, especially the massif of Mauna Loa; then come extinct volcanoes, and in the west coral islands and atolls are built on top of ancient volcanoes which have been gradually sinking.

A few other oceanic islands, like the Seychelles, appear from their rock to belong to the continents. This type is very rare, and their study may throw a great deal of light on the origin of the continental masses.

Volcanoes

These are truly infamous: the eruption of Tambora in Indonesia in 1815 killed 92,000 people. Considerable quantities of energy are involved in a volcanic eruption, often surpassing the power of an atomic explosion (table 8). Active volcanoes in populated regions are kept under constant surveillance by observatories equipped with apparatus to register earth movements, vibrations due to sound, underground temperatures, and so on. Sometimes it is possible to deflect a lava flow by digging a trench or by exploding a bomb in its path, but no way has yet been found to stop an eruption.

The volume of matter thrown out in a single eruption can be as much as 150 cubic kilometres (table 9), and as eruptions may follow one another for millions of years this could be increased a thousand times. The total volume of the largest island of the Hawaiian group, measured from the ocean floor, is more than 100,000 cubic kilometres; that of Iceland is 120,000. The lavas of the Columbia Plateau in the United States and the Deccan Plateau of India occupy more than 300,000 cubic kilometres. Many of the tallest mountains are

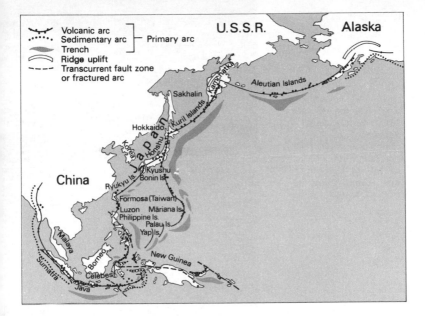

volcanoes; Aconcagua in Chile is 6,959 metres high. Mauna Loa in Hawaii rises over 9,000 metres from the sea floor to its summit 4,200 metres above the surface of the sea, and would rival Everest if its foothills could be seen beneath the sea.

There can be eruptions under the sea as well as on land. If this happens in shallow water jets of vapour, liquid and solid matter can be seen at the surface. When the depth is greater than two thousand four hundred metres, however, the pressure due to the overlying water exceeds its critical pressure and the sea water cannot boil; in this case any gases emitted dissolve, and the lava simply spreads out over the ocean floor with no sign at the surface. Later dredging or boring may gather some of the cooled and solidified lava.

Contrary to what has been said in the older books, the expanses of oceanic floor between two thousand and six thousand metres deep form the main areas of volcanism on the earth. Sheets of lava alternate with layers of mud, and the depths are scattered with submarine mountains in the shape

42

Figure 7. Great areas are covered by 'plateau lavas'. In this photograph the columnar jointing of the lava flows is revealed in the cliffs on the north coast of Ireland near the famous Giant's Causeway.

(1) Fissure or Icelandic type

(2) Hawaiian type

(5) Plinean type

(3) Strombolian type

(6) Pelean type

(4) Vulcanian type

Figures 8 and 9. There are many types of volcanic eruption, and some of the distinctive types are shown on the left. (1) Fissure eruptions are not violent but very large amounts of lava often erupt; some of the Icelandic eruptions were of this type. (2) The Hawaiian type of eruption is also relatively quiet; lava lakes are often formed, from which fountains of red hot lava may sometimes spray. (3) Strombolian eruptions are spasmodic: trapped gases gather below the lava, and masses of lava and ash are periodically blown into the air. (4) The Vulcanian type of eruption is more violent because the more viscous lava solidifies between eruptions and the trapped gases attain a high pressure before the overlying lava is blown out of the crater. (5) The Plinian type of eruption is very violent – lava saturated with gas is blown to a great height and showers of ash descend on the neighbourhood of the volcano: such an eruption by Vesuvius destroyed Pompeii in AD 79 (6) The Peléan type of eruption is one of the most destructive: clouds of incandescent lava droplets and gas roll down the side of the volcano and destroy everything in their path. In May 1902 the city of St Pierre in Martinique was destroyed in a few minutes by the eruption of Mt Pelée, which killed all 30,000 inhabitants. *Below* One of the largest areas of lava in the world is the Columbia Plateau in the north-western United States.

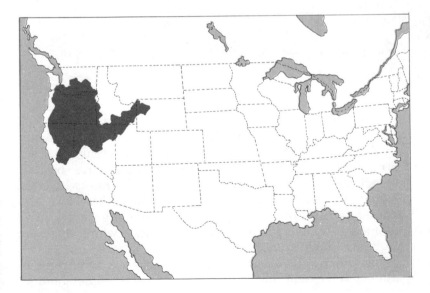

Table 8 Energy liberated in some large volcanic eruptions

Total energy

Volcano	Date	Energy (ergs $\times$ 10^{19})
Tambora (Indonesia)	1815	840,000,000
Sakurazime (Japan)	1914	4,600,000
Bezymianny (Kamchatka)	1956	2,200,000
Krakatoa (Indonesia)	1883	1,000,000
Asama (Japan)	1783	880,000
Fujiyama (Japan)	1707	710,000
Bandaisan (Japan)	1888	10,000
Hydrogen bomb for comparison		200,000

Energy of explosion

Volcano	Date	Energy
Krakatoa	1883	86,000
Bezymianny	1956	36,000
Mont Pelée (Antilles)	1902	470
Asama	1938	17
Vesuvius (Italy)	1944	6
Atom bomb for comparison		140

of cones or truncated cones of ancient volcanoes; the rock dredged up is characteristically basalt lava. Volcanic islands are formed from those volcanoes which have risen above sea level, and coral islands from those which have become extinct and are under water. Around California, where the Pacific Ocean has been most fully explored, there is a submarine volcano over one thousand metres high every forty kilometres on average. Over the globe as a whole there are probably 10,000 or more.

The other volcanic areas are certainly less important in number and extent, but very interesting for an understanding

Table 9 Volume of matter ejected during the greatest volcanic eruptions known

Volcano	Country	Date	Volume in cubic kilometres	
			Lava	Ash etc.
Tambora	Indonesia	1815		150
Veidivatna	Iceland	before 870	43	
Katmai	Alaska	1912		30
Krakatoa	Indonesia	1883		18
Laki	Iceland	1783	12	
Eldgja	Iceland	930	9	
Bezymianny	Kamchatka	1956	3	
Bandaisan	Japan	1888		1
The total from major eruptions		1815–1964	**67**	**199**

Table 10 Number of volcanoes active on land

	HIST	HIST+SOLF
Oceanic islands in the Pacific	17	23
Atlantic	?	40
Indian	?	4
Antarctic volcanic belt		11
Circum-Pacific belt	245 at least	426
Mesogeic zone (Antilles-Mediterranean-Sunda)	80	135
Africa	18	46
Continental Asia	0	5
Total	**360** at least	**690**

HIST = having erupted at least once in historic times
SOLF = having reduced activity : solfataras (emission of steam), etc.

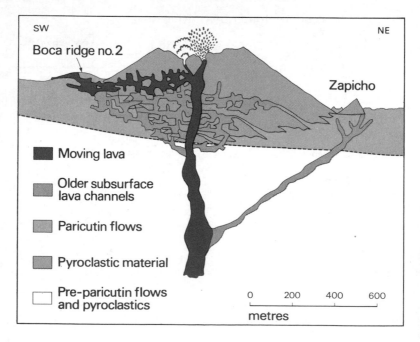

SW

Boca ridge no. 2

NE

Zapicho

Moving lava

Older subsurface lava channels

Paricutin flows

Pyroclastic material

Pre-paricutin flows and pyroclastics

0   200   400   600

metres

of the general structure of the earth (table 10). The most important is the marginal strip between continent and ocean, that belt of fire which surrounds the Pacific, and the Mesogeic zone which extends from the Antilles to Indonesia via the Mediterranean. Towards the interior of the continents a few huge volcanic massifs mark a linear series of fractures from the Lebanon to the Red Sea and eastern Africa, and others are scattered over the heart of Africa (Hoggar, Tibesti) and Asia. Volcanism is clearly less frequent in the continental interiors than at the edge, and here, in turn, it is less frequent than in the depth of the oceans. Volcanoes are often abundant along mountain fold belts, as in Indonesia, the Andes and the Antarctic, but are sometimes rare or marginal, as in the Alps.

We have seen that volcanoes are frequent along island arcs, and the greater the curvature of the arc, the closer the volcanoes are to one another. An idea which immediately comes to mind is that the greater the intensity of crustal deformation the more likely plutonic material is to be erupted. Plutonic material is formed by solidification of molten magma inside the earth and is coarsely crystalline. In fact, volcanoes do often accompany fractures in the crust.

Several new volcanoes have been observed within historic times. One of the most famous is Paricutin in Mexico, which started life in 1943 in a quiet field. Within a week it had built up to a cone of cinders 150 metres high, and it continued to grow by the eruption of ashes, joined later by flows of lava. More recently a new volcanic island has appeared near the coast of Iceland, and has been named Surtsey. It does not take long for an active volcano to reach a great height. Izalco in El Salvador has reached a height of 2,000 metres in less than 200 years since its first appearance. Submarine

Figure 12. The world's active volcanoes: there is a very marked concentration in certain regions, and the greatest concentration forms a belt encircling the Pacific Ocean.

51

volcanoes do not grow so easily because loose ashes and cinders are easily washed away by the waves, and many of them rise to just below the surface of the sea.

Lava sheets spread out from *fissures*, but it is material ascending through vertical *vents* that builds up the familiar volcanic cones. Frequently they contain as high a proportion of ash and cinders as they do of lava. Those vast elliptical flat-bottomed depressions, called *calderas*, are the result of explosions or subsidence, or of some combined phenomenon (see table 11).

Meteorite craters

It would have seemed unbelievable two centuries ago that projectiles from space could be bombarding the earth, gouging out craters hundreds or thousands of metres across. The evidence, however, is inescapable: masses of iron or stone, called *meteorites*, do fall to earth from the sky. The larger ones cause big craters and the heat set free is often sufficiently great to volatilise the meteorite. A meteorite which fell in central Siberia in June 1908 landed with such an impact that trees were destroyed for a distance of 20 kilometres, and the blast was recorded on instruments as far away as the British Isles. Fortunately such falls are very rare.

How can one distinguish between craters caused by meteorites and those due to a volcanic explosion or to subsidence, between craters of cosmic and of terrestrial origin? Meteorite craters are generally more circular and less elongated than others (table 11). Their rocky sides often display signs of plucking or tearing, and two unusual minerals, *stishovite* and *coesite*, which occur at the heart of the craters, are known to form only at high pressures, and

Table 11 Volcanic calderas and meteorite craters

Volcanic calderas

Volcanic calderas	D	d	D:d
1 Buldir (Aleutians, N. America)	43	21	2·05
2 Valles mountains (New Mexico, N. America)	29	25	1·16
3 Asosan (Japan, Asia)	25	16	1·47
4 Aira (Japan, Asia)	23	17	1·38
5 Kikai (Japan, Asia)	23	16	1·44
6 Ata (Japan, Asia)	25	12	2·08
7 Targo Yega (Tibesti, Africa)	20	17	1·18
8 Kawah Indien (Java, Indonesia)	20	16	1·25
9 Conca di Bolsena (Italy, Europe)	17	17	1·00
10 Tenerife (Canary Islands)	20	13	1·54
11 Tarso Voon (Tibesti, Africa)	18	14	1·29
median value	**23**	**17**	**1·38**

D = the maximum diameter in kilometres.
d = the diameter perpendicular to in kilometres.

Meteorite craters (after Preuss and Monod)

Meteorite craters	D	A	F	V	C	S
1 Ries (Bavaria, Germany)	20	A	F	V	C	S
2 Deep Bay (Canada, N. America)	13	—	—	—	—	—
3 Bosumtwi (Ghana, Africa)	11	A	—	V	C	—
4 Serpent Mount (Ohio, N. America)	6·4	A	—	—	C	—
5 Brent Crater (Ontario, Canada)	3·5	—	—	V	—	—
6 Chubb Crater (Ungava, Canada)	3·4	—	—	—	—	—
7 Köfels (Otztal, Austrian Tyrol)	3·0	—	—	V	—	—
8 Hollesford (Ontario, Canada)	2·3	—	—	V	C	—
9 Tenoumer (Mauritania)	1·8	—	—	—	—	—
10 Barringer Crater (Arizona, U.S.A.)	1·2	A	F	V	C	S
median value	**3·5**					

A = marks of plucking   F = meteoritic iron   V = glass (tektite)
C = coesite   S = stishovite

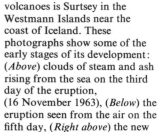

Figure 13. One of the newest volcanoes is Surtsey in the Westmann Islands near the coast of Iceland. These photographs show some of the early stages of its development: (*Above*) clouds of steam and ash rising from the sea on the third day of the eruption, (16 November 1963), (*Below*) the eruption seen from the air on the fifth day, (*Right above*) the new volcano, now 70 metres high, on the eighth day of the eruption, (*Right below*) an aerial view of the island on the seventeenth day.

have also been found at the sites of atomic explosions. The meteoritic origin of a crater is clearly diagnosed if one can find the remains of the meteorite itself, as in the great meteor crater of Arizona, where it did not wholly volatilise.

Our neighbours the moon and Mars display many similar craters, clearly visible on the photographs taken at close range from space craft. They are moreover more numerous and better preserved than upon earth, because craters on the earth stand a good chance of being obliterated by erosion or masked by other phenomena, in contrast to the moon where such effects are almost negligible.

Ice sheets

In addition to rock and water, the third constituent of the earth in contact with the atmosphere is ice. If we find the beauty and grandeur of mountain glaciers impressive, in what terms shall we describe the immense dome of ice which covers almost the entire Antarctic continent? It has a surface area of 12,160,000 square kilometres, is bigger than Europe, and its average thickness is 1,850 to 2,200 metres, and it can reach as much as 4,700 metres.

In all, the world's glaciers and ice caps occupy about 14,500,000 square kilometres, which is three per cent of the earth's surface and eight per cent of that of the continents (table 12). They have a volume of twenty-six million cubic kilometres, the equivalent of twenty-four million cubic kilometres of water, or 130 times more than the volume of terrestrial fresh water. If they were to melt, the sea level would be raised by sixty metres throughout the world, submerging New York, Paris, London, Barcelona, Rome and Cologne!

The two great continental ice caps of Antarctica and

Figure 14. This view of Ben Nevis, Scotland from the north-east shows cliffs of andesite lavas and breccias, and shows the contrast between the ice-moulded bottom of the valley and the frost shattered cliffs above. Ice and frost are two of the most important agents that shape the surface of the earth.

Figures 15 and 16. Below. The formation of a caldera. The top diagram shows a cross-section through an active volcano underlain by a body of magma, some of which has erupted at the surface as lava. Weakened by the eruption of some of the underlying magma, the cone collapses into the magma chambers leaving a

Greenland alone comprise more than ninety-seven per cent of the area of land ice, and more than ninety-nine per cent of its volume. Their unique dome-like form, once acquired, is self-perpetuating because the ice flowing slowly to the edges is replaced in the centre by snow, which does not melt at the high altitudes and settles down, changing in time to ice.

The excess weight of the big ice sheets may cause the basement rock to sink. We are aware of this because Scandinavia,

huge depression at the surface of the ground, as shown in
the lower diagram. *Below*. Crater Lake, Oregon, with Wizard Isle
in the foreground. Crater Lake is an example of a
caldera, formed by the collapse of a former volcano.

Table 12 The largest ice caps (after Corbel, Kalesnik and Cailleux)

Continental ice domes	Lowest Ice temperature	Area in km²	R
Antarctica	−50	13,100,000	7·6
Greenland	−20	1,726,000	69·04
Local ice caps			
Northern Ellesmere (Canada)	−20	25,640	1·17
Central Ellesmere (Canada)	−20	22,000	1·05
East Spitzbergen	−15	21,000	1·05
Southern Ellesmere (Canada)	−20	20,000	1·32
Devon Island (Canada)	−15	15,125	1·12
Southern Patagonia	−10	13,500	1·23
North-east Spitsbergen	−15	11,000	1·18
Axel Heiberg (Canada)	−20	9,300	1·06
Vatnajökull (Iceland)	− 5	8,800	1·45
Barnes, Baffin Island (Canada)	−15	6,050	

R = ratio between surface area of one
ice sheet and next on list.

which once carried such a mass of ice, began to rise again when the ice melted and is still doing so, at the rate of a metre per century in some places. The great depth (500 metres) of the continental shelf round Antarctica and Greenland would be explained by the weight of the neighbouring huge ice cap.

Oceans

Each ocean is composed of three parts: the continental shelf, which we have seen is an extension of the continent itself, the continental slopes, and the great depths beyond. The continental shelf dips slightly, about two metres per kilo-

Table 13 The largest seas

		Area in km^2	Average depth in metres	Maximum depth in metres	Neighbouring ocean
1	Caribbean Sea *	2,750,000	2,490	7,491	Atlantic
2	Mediterranean Sea	2,500,000	1,485	4,901	Atlantic
3	Bering Sea *	2,270,000	1,436	3,961	Pacific
4	Gulf of Mexico	1,540,000	1,512	4,376	Atlantic
5	Sea of Okhotsk *	1,530,000	838	3,379	Pacific
6	East China Sea *	1,250,000	188	2,681	Pacific
7	Hudson Bay	1,230,000	128	258	Atlantic
8	Sea of Japan *	1,010,000	1,350	3,617	Pacific
9	Andaman Sea *	800,000	870	4,177	Indian
10	North Sea	580,000	94	725	Atlantic
11	Black Sea	460,000	1,100	2,246	Atlantic
12	Red Sea	440,000	491	2,359	Indian
13	Baltic Sea	420,000	56	459	Atlantic
14	Persian Gulf	194,000	30	84	Indian

*Seas which are divided from the open ocean by an island arc.

metre, and its outer limit against the continental slope is always well-defined but at a depth varying between twenty and five hundred metres, with its mean at 130 metres. The continuity between the coastal plains and the continental shelves is remarkable, and has been attributed to erosion during a period of lowered sea level. R. A. Daly has estimated how much the sea level may have been lowered during the glacial periods due to the accumulation of ice on land; most recent estimates are as much as 75 or even perhaps up to 140 metres. It may be that the smoothness of the continental shelf can be explained in part by wave action down to that depth during glacial times.

The continental slope goes down more rapidly, from sixty

Figure 17. Some typical volcanic phenomena shown by the new volcano Surtsey: *Far left*. A lava lake in the crater, emitting a continuous fountain of incandescent lava. *Left bottom*. A lava flow entering the sea on the south-west side of the new island. *Left*. Lava overflowing from the crater and escaping to the sea – steam is rising from the point at which the lava enters the water. *Below*. An explosive phase of the eruption.

Figure 18. Meteor Crater, Arizona, looking westwards. This is the best known of several large craters formed by the impact of giant meteorites. High-pressure minerals found in the floor of the crater testify to its violent mode of origin.

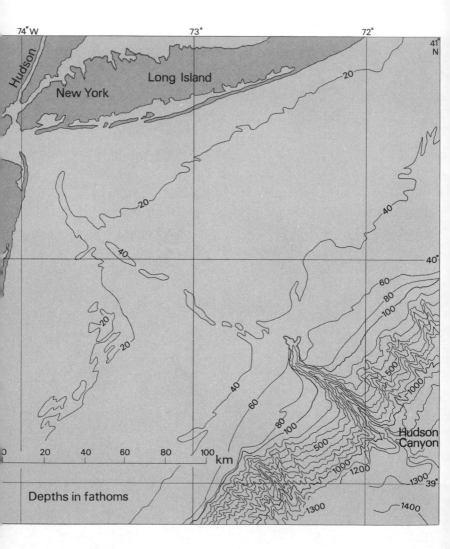

Figure 19. The Hudson Canyon, one of the submarine canyons which crosses the continental shelf. Some of them, like this one, appear to be located off the mouths of large rivers.

Depths in fathoms

to seventy metres per kilometre to three thousand or four thousand metres. It is grooved by relatively steep-sided canyons two thousand or even three thousand metres deep. How could these canyons have been gouged out of the hard rock? The experts are not in agreement, but some believe that they were cut by submarine currents loaded with debris from the continental shelves. At any rate, they are partly filled by fine grained sediment which, being unstable, tends to slump, and the muddy water being more dense flows down the slope and the mud is deposited farther out to form fans of sediment at the foot of the canyons. Some of the canyons lie off the mouths of large rivers or the continuations of their valleys on the continental shelf, notably those off the River Hudson and the Congo. Others seem to be independent of recent valleys, for instance, those in the Atlantic between Arcachon and Ireland.

The great depths below and slightly above four thousand metres occupy three quarters of the Pacific and a third of the Atlantic and Indian Oceans. The remarkable *sea mounts* – the remains of ancient volcanoes – are sporadically distributed across the sea floor. Apart from these the floors of the deep oceans consist of huge abyssal plains, with broad ridges and valleys in the centres of the oceans. The great *abyssal plains* are almost horizontal, being covered by the fine deposits of mud carried out by currents; the irregularities hardly exceed a gradient of one or ten metres in a thousand. Apart from the sea mounts, there are more modest abyssal hills scattered about, from ten to a hundred metres high with a width ten to forty times this, formed of volcanic lavas alternating with mud. The hills are sometimes aligned as though along fissures. Their frequency is fairly constant, unlike the sea mounts which tend to be grouped in distinct clusters.

Figure 20. The eruption of Tristan da Cunha in October 1961 is an example of the volcanicity associated with the mid-oceanic ridges. This picture was taken when H.M.S. Leopard went to the island to collect some of the islanders' belongings, which they had to leave behind during the hasty evacuation.

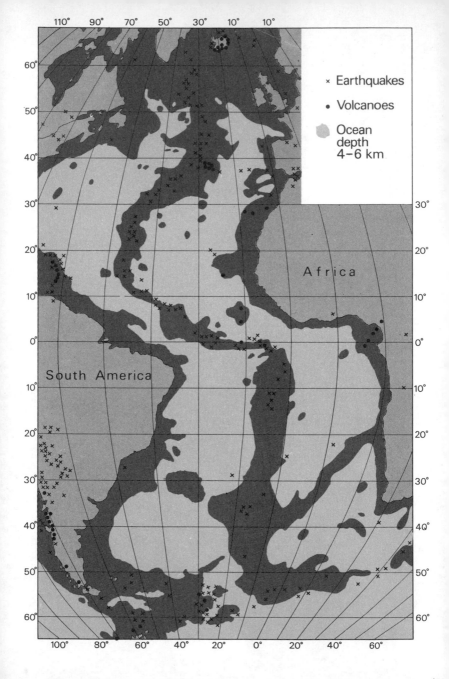

Figure 21. The mid-Atlantic ridge: the submarine mountain chain which extends for thousands of miles, emerging above the surface only occasionally in volcanic islands, such as on the Azores and Ascension Island. Many of the earthquakes in the Atlantic occur along the ridge.

The Atlantic Ocean is divided from north to south at its centre by a *ridge*, as are the Arctic Ocean (by the Lomonossov ridge) and the Indian Ocean (whose ridge bifurcates to the south). Another ridge extends to the centre of the Southern seas and forms an irregular belt which encircles the Antarctic continent some distance from land. The ridges of the Pacific are less clear and their characters more blurred. So few soundings have been made in the more remote parts of the oceans that the exact extent of the mid-ocean ridges is as yet uncertain, but it seems probable that they form a continuous range of submarine mountains, several hundred kilometres wide and 60,000 kilometres in length. The highest points rise thousands of metres from the ocean floor, sometimes breaking above the surface. The volcanic islands of Iceland, the Azores, Ascension and Tristan da Cunha form part of the mid-Atlantic ridge. As well as being regions of volcanic activity the ridges are also marked by belts of shallow earthquakes. These ridges form one of the most fundamental topographic features of the earth's surface, comparable in size to even the largest continental mountain ranges.

The oceanic trenches are something quite different. They cut the ocean floor at four thousand to five thousand metres, and descend several thousand metres further, varying from fifty to one hundred kilometres in width. The slope of their sides is relatively gentle, five to ten per cent. The floor is flat, is probably filled in with mud, and is from 0·5 to 3·0 kilometres wide. The trenches are elongated, curved, situated near the edges of the oceans, and with the single exception of the Diamantina trench, they are always parallel to an island arc, a mountain chain or a line of volcanoes. Thus the greatest depths of the earth are found close to areas of high relief. Table 4 gives a list of the principal trenches. They are mainly concentrated in the Pacific, more rarely in the Atlantic and the Indian Oceans, and are absent in the Arctic, which is relatively shallow compared with the other oceans.

Comparison between oceans and continents

The external forms of the two great domains into which the surface of the earth is divided – the ocean floor with its islands, and the continents with their shelves – have a few points of resemblance and many differences. The principal resemblance lies in their dissected nature, but this does not necessarily mean that both domains were formed by the same forces, for the processes of erosion which shape the continents do not operate in the depths of the oceans.

We have seen that the oceans form one continuous whole and the continents are separate masses. The oceans have ridges situated especially in the middle, the continents have mountains particularly along the coasts. The oceans have deep trenches whereas the continents have not. Volcanoes and lava flows are incomparably more widespread on the ocean floor than upon the continents. There is thus a fundamental contrast between the two domains. What is its origin? How have the distinctive features of the continents and ocean basins arisen? To understand these we must look more closely at the structure of the earth's crust and its interior.

2 The earth's interior

There is very little information that can be obtained directly about the inner structure of the earth. We know from measurements in mines and boreholes that the temperature increases with depth, and we also know from astronomical measurements that the earth's average density is much greater than that of any of the rocks visible at the surface. Nevertheless, there are indirect ways of obtaining information about the interior of the earth, and our principal guide will be the study of earthquakes and artificial explosions.

Information from earthquakes

The word earthquake evokes impressions of abrupt and terrible natural catastrophies. One in Japan in 1923 caused the death of 99,000 people and destroyed 129,000 houses. It is for just this reason that earthquakes have been the subject of detailed and careful study, in which scientists from Japan and other earthquake-prone regions have taken a leading part. Instruments have been developed capable of registering the faintest shock waves from distant earthquakes. The results of this work have been so interesting to investigators of the earth's structure, as we shall see, that use is now made of artificial explosions, which have the added advantage of being predictable.

The shock caused by an earthquake travels in waves, of which there are three kinds: the primary (P) waves, so called because they arrive first at the place of recording, the secondary (S) waves which arrive a few minutes later, and finally the main surface (L) waves, which are even slower. The speed with which they are propagated varies with the nature of the material through which they are passing; table 14 gives some examples for primary waves. By calculating the

Figure 22. The record of an earthquake in Turkey made by a seismograph at Pulkovo in Russia, 2,200 kilometres away. The distance is known from the time interval between the arrival of P and S waves (3 minutes 43 seconds).

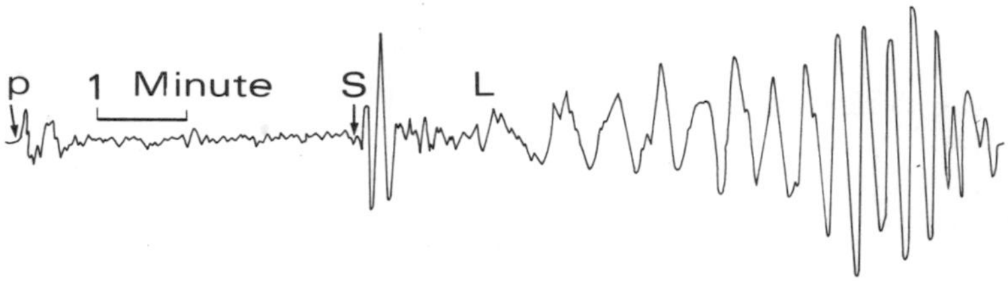

speeds of propagation one may get some idea of what sort of materials the waves have passed through in the earth.

From the times of arrival of the primary waves at different observatories, it is possible to calculate the position of the *focus* of the tremor, the point where the shock originated, even if it were under the sea or in an inaccessible region. This gives us the location of the point on the earth's surface directly above the focus, which is called the *epicentre*. One can also obtain some idea of the depth of the earthquake.

Like all waves when they encounter a different medium, earthquake waves may be reflected, that is to say, sent back like an echo, or they may be refracted, which means that they change direction slightly on passing from one medium into another. Both of these properties enable us to measure the thickness of the various layers which they have crossed. The surface waves, so called because they are propagated only in the crust, also provide a measure of the thickness of the crust, thanks to the dispersion of their long wave lengths. From the amplitude of the oscillations received on the seismograph which registers the shock it is possible to calculate the energy of the earthquake at its focus, and thus classify it according to a scale of 'magnitudes'.

The most interesting result of these observations is that the seismograph recordings are always very similar, no

Table 14 Propagation velocity of primary (P) earthquake waves
in kilometres per second

	Minimum	Mean	Maximum
Sea water		1·5	
Gravel, sand, mud	0·5		2·2
Schist	1·5		4·5
Calcite	3·0		5·6
Dolomite	4·9		6·0
Granite	3·9		6·3
Basalt	4·8		6·0
Quartzite		6·1	
Marble		6·5	
Gabbro	6·9		7·2
Peridotite	7·2		8·2

matter where the earthquake occurs or at which observatories it is recorded. In other words the earth as a whole must have the same properties in all directions, consisting essentially of a series of uniform concentric shells, disregarding superficial irregularities.

Our knowledge of the earth's internal structure is largely derived from a study of the P and S earthquake waves. In general, the velocities of these waves increase with depth, but it has been found that there is a sharp jump in the earthquake wave velocities at two particular levels. One of these is called the Mohorovičić discontinuity, after its discoverer, and is at a depth of about 30 to 40 kilometres below the continents, and less under the ocean floors. The other occurs at a depth of 2,900 kilometres, about half way to the centre of the earth. These discontinuities represent major changes in the composition or physical properties of the material through which the waves pass, and they serve to divide the earth into three parts: the *crust*, a relatively thin surface layer extending from the surface down to the Mohorovičić discontinuity; the *mantle*, extending down to 2,900 kilometres; and the *core*, extending down to the centre of the earth. Table 15 summarises the main subdivisions of the earth.

The P and S waves differ in an important respect; the former are compressional waves, and can travel through solids or liquids, but the latter are distortional waves and can only travel through solids. By plotting the paths of waves from distant earthquakes it has been found that S waves do not pass through the core of the earth, and from this it is inferred that the material of the core is a liquid. For example, in the British Isles only the P waves would be detected from an earthquake taking place in New Zealand because the core prevents S waves from passing through.

The core itself may be divided into an inner and an outer part, but seismologists are not yet certain whether the inner core has a sharp boundary or a transitional one. It is not even certain whether the inner core is liquid, but since the outer core will not transmit S waves we cannot use this method of finding out.

One might expect that the transmission of S waves would also be hindered by molten rock in the crust and mantle, since the widespread distribution of volcanoes indicates that lava must be present under the surface in many areas. In fact this is not so, and melting is evidently restricted to small pockets situated under active volcanoes.

Apart from the valuable information which can be derived from them, earthquakes are of great interest in themselves. The energy which is involved is enormous. The greatest recorded since 1906 was the one in eastern Tibet in 1950. It had a magnitude of 8·6, equivalent to 100,000 atom bombs

Table 15 The main subdivisions of the earth. (After Bartels, Brinkman, Bullen, Dauvillier, Jeffreys, Leet, Scheidegger, Sitter and Strahler.)

Depth in kilometres	Discontinuity	Layer	Possible chemical composition
		Upper crust	Granite
17–25	Conrad		
		Lower crust	Basalt? Gabbro?
32–38	Mohorovičić		
		Mantle	Peridotite?
2,900	Gutenberg		
		Outer core	Ferro-nickel?
5,000–5,200	Lehmann		
		Inner core	Ferro-nickel?
5,371			

Velocity of primary waves in km/sec	State	Density	Temperature in °c
5·5-6·1	Solid	2·7	
			400
6·4-7·2	Solid	3·0	
			600-1,000
8·0-8·2	Solid	3·3	
13·6		5·3-6·7	
			1,500-5,500
8·1	Liquid	9·0-10·5	
9·4-10·4		11·5	
11·2-11·7		11-17	
			1,900-6,000
	Solid		
11·2-11·7		12-18	1,900-10,000

Table 16 Number of earthquakes at shallow depths per year from 1904-1946 after Gutenberg

	Magnitude	Number
Very important	8·6 − 7·7	2
Important	7·7 − 7·0	12
Fairly important	7·0 − 6·0	108
	6·0 − 5·0	800
	5·0 − 4·0	6,200
Significant	4·0 − 3·0	49,000
Felt by man	3·0 − 2·5	100,000
	etc.	etc.

or 100 hydrogen bombs. To judge from the descriptions, the terrible shock which destroyed Lisbon in 1755 must have reached a magnitude of 8·7 or 9, and no earthquake is known to have exceeded this.

Each year there are 150,000 earthquakes strong enough to be felt somewhere on earth. From table 16 it may be seen that happily, the stronger ones are the rarest. The most violent shocks are responsible for by far the greatest part of the total energy liberated each year. The foci of the earthquakes range from the surface down to about 700 kilometres. They may thus originate not only in the crust but also deep in the mantle.

We know that the greatest visible earth tremors are associated with planar fractures in the crust, called *faults*. The rocks on either side of a fault are displaced relative to one another, sometimes by as much as 100 kilometres, but more commonly only a few metres or hundreds of metres. Individual movements on fault planes rarely exceed a metre or two, even in large earthquakes, and the great displacements on large faults take place by repeated movement over thousands of years. The great earthquake which destroyed San Francisco in 1906 was caused by a movement of five metres on the San Andreas Fault, which passes within 10

Figure 23. A fault near San Francisco due to an earthquake.

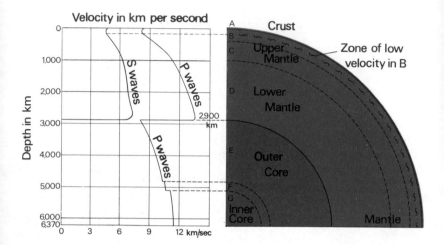

Figure 24. The major subdivisions of the earth can be distinguished by the difference in velocity of seismic waves passing through them. The outer core does not transmit S waves and hence can be inferred to be made of a liquid material. The core and mantle can be distinguished most clearly, but other minor seismic discontinuities can be detected as well.

kilometres of the centre of the city. Many faults have been found by geologists in regions which are no longer subject to earthquakes, although these must have occured in the past.

Looking at a world map of earthquake foci, and taking into account only the most intense, that is, those with a magnitude greater than 5·5, we find that certain regions are much more subject to earthquakes than others. The foci are all situated in regions which are considered unstable for other reasons, and especially in mountain belts of geologically recent origin and their surroundings, and 80 per cent of the energy is liberated from the circum-Pacific belt. Fifteen per cent comes from the Mesogeic zone (Antilles-Mediterranean-Himalayas). The remaining 5 per cent is scattered over the rest of the earth, still showing some striking areas of concentration, notably the crests of the mid-oceanic ridges. There

Figure 25. Cross sections of the earth showing the paths of seismic waves arising from an earthquake. The P waves can reach the far side of the earth in approximately 20 minutes, but there is a shadow zone where no waves are recorded because of refraction at the core-mantle boundary.

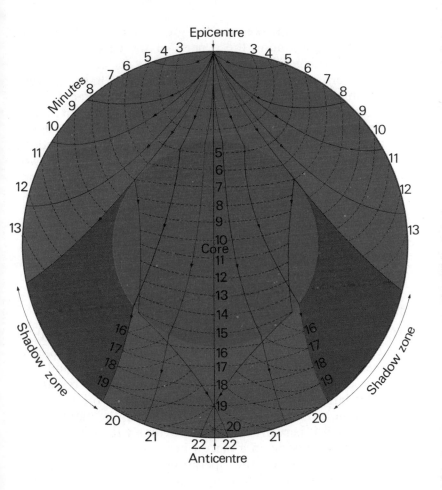

are very few earthquakes on the great plains of the ocean floor. These are among those parts of the earth which are considered with good reason to be the most stable.

The intermediate and deep earthquake foci have an even more limited distribution, occuring in the circum-Pacific belt, the Mesogeic zone, and the Scotia arc situated between South America and the Antarctic continent. They mark the island arcs and curved mountain ranges of Mexico and the Andes, underlining once again the relationship between these two features. The foci always follow the same pattern as one approaches the arcs from the open ocean towards the continent. First come the shallow earthquakes, then the intermediate ones, and finally the deepest. This leaves a wedge of rock in which there are relatively few earthquake foci, bounded by a surface dipping at ten to forty-five degrees towards the continent, along which there are many.

The crust

The crust is not only the most accessible part of the earth's structure but also the most complex. The crust is much thinner under the oceans than under the continents, and the oceanic crust has been the subject of much recent research. This has largely been inspired by the Mohole project, a plan to drill a hole deep enough to penetrate into the mantle and find out what it is really made of. Obviously if such a hole is ever drilled it must be in a place where the crust is very thin, otherwise the cost of the operation would be too great. Despite what is written in many books, the crust is much the same under all the oceans. Certainly each specialist has a tendency to emphasise particular differences in composition which he has succeeded in revealing, but this does not alter the

Table 17 Structure of the ocean floor (after Raitt)

	Thickness in kilometres	Velocity of P waves in km/sec.
Sea water	4·5	1·5
1 Unconsolidated sediment	0·3 ± 0·1	2·0 ± 0·2
2 Sediments and lavas ?	1·7 ± 0·8	5·1 ± 0·6
3 Basalt or gabbro	4·9 ± 1·4	6·7 ± 0·3
Rounded sum of 1 + 2 + 3	7	
4 Mantle (peridotite ?)	2,900	8·1 ± 0·2

The figures given here apply only to the deep ocean floors, not to the continental slopes, mid-oceanic ridges, flanks of islands, or oceanic trenches.

fact that such differences are minimal, and are of the same order as the differences in average depth between the three oceans.

In the great basins under 3,000 to 7,000 metres of water that are far from continents, islands, or mid-oceanic ridges, one first encounters a layer of mud and other soft deposits, which is about 300 metres thick on average. In certain places on the surface are scattered nodules of calcium phosphate and manganese oxide, which in some places are almost abundant enough to repay commercial exploitation, and into which research is at present being conducted. Under layer 1 (table 17) there comes an intermediate layer 2 about 1,700 metres thick, probably formed of alternating mud, more or less consolidated, and sheets of volcanic lava extruded under water. Then comes the principal layer 3, having the physical properties of basalt or gabbro and with a thickness of about 5,000 metres. Under this comes the mantle-layer 4.

Under atolls and other coral islands the deep structure (layers 2, 3 and 4) is the same, but in place of three hundred

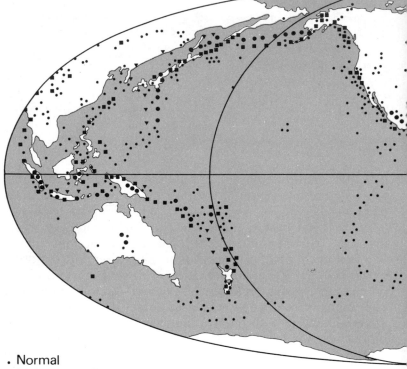

Figure 26. World distribution of earthquakes. The regions which are prone to earthquakes are often also the site of volcanoes (see figure 12). A major earthquake belt encircles the Pacific and another extends eastwards from the Mediterranean towards the Himalayas and the East Indies. The mid-oceanic ridges are also the site of many earthquakes.

- Normal
- Major superficial
- Major intermediate
- Major deep

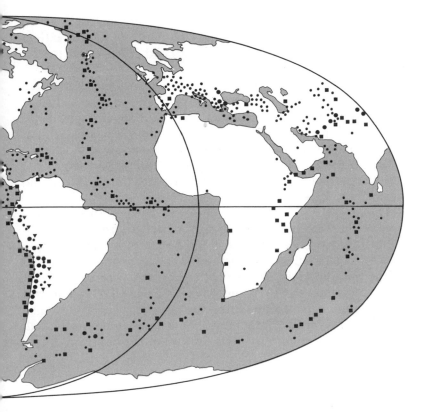

metres of mud and 4,700 metres of water one finds about 4,000 to 5,500 metres of basalt (the material of the extinct volcano on which the island is based) and 500 to 2,000 metres of coral limestone.

Close to the continents the thickness of soft unconsolidated deposits (layer 1) is much greater, as might be expected since it is formed directly or indirectly of sediment brought down by rivers from the neighbouring land. In the basins around California this thickness reaches 3,000 metres, and more than 6,100 on the continental shelf in the Gulf of Mexico, 5,200 on the continental shelf along the eastern United States, and as much as 9,200 on the continental slope and at its foot.

Under the continents the thickness of the crust is much greater than in oceanic areas and its structure is more complex. In most places the crust can be divided into an upper and a lower part, of which the lower has similar properties to the basaltic layer underlying the oceans. The upper part, from the velocities of seismic waves, is believed to be made of rocks similar in composition to granite. There are also sedimentary rocks, both consolidated and unconsolidated, distributed irregularly over the surface of the continents.

Equilibrium of crust and mantle

It is extremely useful to know the intensity of the gravitational force over the surface of the earth, even if it is only to assist in prospecting for subterranean mineral deposits. Above a deposit of heavy minerals the intensity is greater, and over a dome of salt, which is a very light mineral, the intensity is much weaker. For this reason among others, instruments called *gravimeters* have been developed which can measure the intensity with an accuracy of one part in a hundred million. Variations in the strength of gravity are expressed in milligals, the milligal being one-thousandth of a gal (named in honour of Galileo), and the average strength of gravity at sea level is about 980 gals. The measured intensity at any point is usually compared with the theoretical intensity due to a sphere, or more exactly, an ellipsoid with its surface representing as far as possible sea level, and the difference between the measured and the theoretical values is called a gravity *anomaly*. Before making the comparison, however, it is necessary to apply several corrections to the measured values. An obvious correction is to allow for the decreased gravitational intensity due to the height of the point of measurement above sea level, which is about 0·3 milligals per metre. This altitude correction is called the *free-air* correction because it is based on the assumption that there is nothing present between the point of measurement and sea level. In nature, however, there is no air between the two points, but rock, and this mass of rock exerts a significant and calculable force of attraction at the point of measurement, which also necessitates a correction. The difference between observed and calculated intensities of gravity remaining after these two corrections have been made is called the *Bouguer* anomaly, after a French geodesist who in the eighteenth century was the first to realise the significance of variations in gravity.

The measurement of Bouguer anomalies at various points over land and sea led immediately to a fundamental result. The anomaly is almost always negative over land, and positive over the ocean floor, and it is numerically greater the greater the distance from sea level. Putting this another way, the high areas of the earth's surface must be underlain by light material, and the low areas by heavy material. Thus the

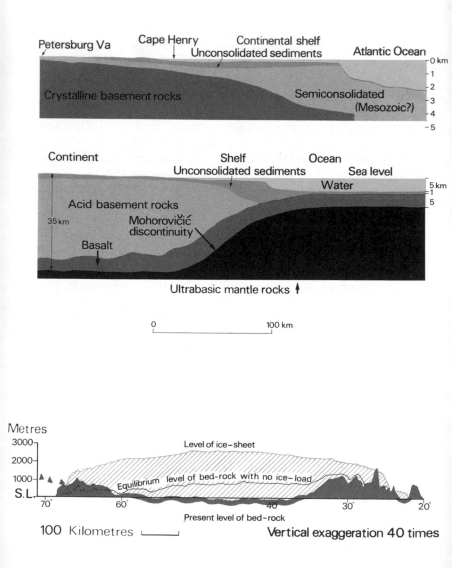

Figure 27. **Top** generalised sections across a continental margin of the eastern coast of the United States. A great thickness of sediments has accumulated at the edge of the continent, where material brought down by rivers has built up in the sea. **Bottom** section across the Greenland ice-sheet. The weight of ice has depressed the bed-rock to below sea level. If the ice vanished, the land would slowly rise to above sea level.

Figure 28. Wooden blocks of different thickness floating in a tank of water can be used to illustrate the principle of isostasy, and may be compared with the light rocks of the earth's crust resting on the heavier rocks of the mantle. There are several ways in which isostatic equilibrium can be achieved, depending on the distribution of rocks of different density within the crust. This interpretation is due to the English geophysicist Airy.

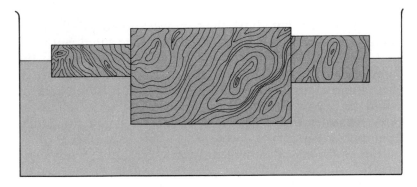

different parts into which one could mentally cut the earth's crust to form vertical-sided blocks are generally in equilibrium. This is described as *isostasy* (from the Greek *stas* meaning *state*, and *isos* meaning *equal*). The various parts of the crust are, as it were, floating upon the substance beneath, like rafts on water. There is a depth below which pressure is equalised as it would be in a liquid, and this is called the *compensation depth*.

An immediate objection springs to mind. How can the substance beneath the crust behave like a liquid when earthquakes and explosions demonstrate that it does in fact have the properties of a firm and elastic solid. The theory of mechanics and our experience will show that the two facts are compatible. The same body can act as a solid towards short sharp forces, such as earth tremors, which last a few seconds, while acting as a liquid or a paste towards more persistent and slow constraints taking place over the long course of geological time. One may compare it with an extremely viscous liquid such as sealing-wax, which breaks like a solid under a sharp blow but bends along its length if it is left unsupported in a drawer.

We have seen that above the compensation depth the pressures due to the various parts of the crust are equal, but these pressures depend upon the magnitude of two factors: the thickness of the piece of crust, and its density. There are in theory an infinite number of ways in which isostatic balance could be achieved; for example the English geophysicist Airy suggested that a light skin or crust is floating in equilibrium on a denser substratum, and where its surface is low it is thin, and where it is high it is thick. It is quite remarkable that this should be the same picture to which we were led by the study of earthquakes: a light crust with a density of 2·7 to 2·8 resting on the heavier mantle with a density of about 3·3. The crust is only five to seven kilometres thick under the depressed ocean floor, of medium thickness (thirty-seven kilometres) under the continental plains, but as much as forty to seventy kilometres under the mountains, where it forms a sort of root. A calculation confirms that the pressure due to these three main types of crustal structure is the same at the depth of 100 kilometres or so. It is probable that on a global scale most of isostatic compensation is accomplished in this way, that is, by the play of different thicknesses of crust floating on the mantle. This has not been definitely proved, however.

Recent work has shown that the transition between 6 kilometre thick oceanic crust and 30–35 kilometre thick continental crust takes place under the oceans in that region where the water is about 2,000 metres deep. It is quite remarkable that it should be no lower but just at the level where, as we have seen, the great continental and oceanic blocks still retain the same general configuration that they display at the surface. The actual passage from one sort of crust into the other spans one to two hundred kilometres, which is not very much; it causes the Mohorovičić discontinuity separating crust and mantle to dip fairly steeply under the continents.

One confirmation of the existence of isostasy is provided by the great ice sheets of Greenland and Antarctica. Although the capping of ice is more than three to four kilometres thick, measurements show that it is remarkably accurately compensated for. When a similar ice cap covering Scandinavia melted tens of thousands of years ago, the land began to rise again, a fact which we can tell from the evidence of raised beaches, and it is still slowly rising. Altogether the area has risen by over 500 metres, and isostatic readjustment has not yet been achieved. There still remain some 210 metres of height to regain, which threatens to leave a good number of Baltic ports high and dry, although it is true that it takes a long time for all this to happen.

The time already taken for Scandinavia to become readjusted allows one to calculate the viscosity of the heavy material which has caused it to be pushed up from beneath. It works out to be a hundred thousand million million million times that of water. One can hardly imagine any material as viscous as this, but it is of course a solid, and only flows under immense pressure and very slowly.

The area of isostatic compensation is not strictly limited to the actual position of the load on the crust, nor to the area from which it is removed; it extends into the surrounding areas by as much as tens of kilometres. The crust as a whole thus has a certain rigidity. This explains why small irregularities in relief, or masses of rock of abnormally high or low density that are less than twenty or so kilometres wide, are not compensated for in the way described above, for example a salt dome or even a feature as large as the Grand Canyon of Colorado. If the latter were compensated for, its floor would be warped and one could observe this from the outcrops of the strata on either side, but this is not the case and the layers of rock are level. In the curious circular depression of Richat in Mauretania, which is fifty kilometres wide, the strata do dome up and it is possible here that isostatic readjustment has played some part although there are other forces which could be responsible for doming.

To sum up, we may say that all the large and important divisions of the earth which we distinguished in chapter 1 – oceans, continents, ice caps, and the crust which supports them – are in mutual equilibrium with the substance beneath, which is somewhat heavier, and upon which they float like rafts. We know moreover that the continents are being constantly worn away by erosion, which removes on average 53 to 87 metres of rock of average density 2·7 every million years. The oceans become loaded with the corresponding deposits of eroded material, which augment the floor by 22 to 36 metres of equivalent compacted rock in the same period. For the overall equilibrium to be maintained it is necessary for an equal mass of matter (220 to 350 million metric tons a year) to be transported at depth from beneath the oceans to beneath the land. This movement is extremely important; directed from the centre of the oceans towards their borders it has a tendency to pull at their floor, setting up considerable strains. The greatest degree of isostatic inequilibrium must be along the continental margins, because it is here that sedimentation is greatest. It is possible that it plays a part in the formation of island arcs and even mountain-building, although we shall see that deep seated currents of another origin are perhaps a more likely cause according to many authors.

At what depth do the compensating movements take place? Possibly in the mantle, more probably somewhere near the boundary between the crust and mantle. At this depth the rock present would have a density between 2·8 (lower crustal material) and 3·3 (mantle material); let us call it 3·0. If in a million years an average of 22 metres of rock density 2·7 is eroded from the continents, in order to re-establish equilibrium 20 metres of density 3·0 must find its way back underneath. The height of the continent will have gone down by the difference: of two metres. In such a cycle even the highest of the present continents, Euro-Asia, would be levelled in 430 million years, and the earth has been in existence for about ten times this period. There must thus be some additional reverse trend to counter-balance the levelling due to erosion in the course of time: the continents must be re-elevated. Taking isostasy into account, this means that they must have a tendency to become thicker. To what force could this be due? The tension exerted by the compensating movements beneath them, or by other currents? We shall not lose sight of this important question.

Terrestrial heat

One may find numerous calculations in old books of the temperatures in the interior of the earth, but since the discovery of radioactivity the whole question has been re-opened. The greatest natural temperatures which have been measured are 1,350°C in the flames of blisters above lavas in Hawaii and Nyiragongo (Central Africa) and 1,200°C in the lavas themselves. It has long been known that the deeper one goes in a mine the higher the temperature becomes, the average being one degree rise in temperature every thirty

Figure 29. 'Old Faithful', one of about a hundred geysers in Yellowstone National Park. A column of hot water and steam is erupted into the air for a period of about five minutes every hour.
Geysers occur in several volcanic regions and are sometimes used as a source of power or hot water.

metres, with extremes of one every eight and one every 125 metres. Knowing the thermal properties of the various rocks, and the temperature gradients which exist in different places, it is possible to calculate the heat transfer across the surface of the earth's crust at each point. This has a mean value of 1·2 millionths of a calorie per second per square centimetre under the continents, and is of a similar order under the oceans. This is a sufficient flow of heat to melt six or seven millimetres of ice per year. For the whole earth the annual heat loss is about two hundred million million million calories.

For comparison with this continuous outward flow of heat, the loss of energy in volcanic eruptions is a hundred times smaller (about fifty times if one takes submarine volcanism into account), and in earthquakes four hundred times less energy is dissipated. On the other hand, the heat received from the sun is three thousand times greater, and if the earth is not gradually warming up through this effect it is because it radiates the sun's heat back into the sky. The hotter the earth's surface, the greater is the re-emission. Between the gains and losses of heat an equilibrium is established which results in an average surface temperature of about 8°C. In this equilibrium the sun's heat is 3,000 times more important than any internal heat of the earth, which is almost negligible in this connection, and it is the sun which dominates our climate. Nevertheless the internal heat of the earth is still much greater than any amount of heat that man can produce. In fact in Iceland, Tuscany, New Zealand and Kamchatka thermal springs or vapour are used as a source of heat, and people use this energy for electricity or hot water. Conversely, the temperature in deep mines is such that one is forced to spend money on cooling them. Ultimately the flow

of heat from the earth's interior helps to explain how in some places the rocks can melt at depths between ten and a hundred kilometres and feed volcanoes.

There is a choice between many possible sources for the outward heat flow:

1 original heat contained in the particles of matter which formed the earth
2 heat due to gravitation, which would be produced in a young planet which condensed little by little with a consequent decrease in its radius
3 heat produced by radioactivity
4 heat produced by chemical reactions
5 frictional heat resulting from movements within the earth.

Of these five sources, only one has been measured, and that is radioactivity, the most recently discovered. The radioactivity in the earth is principally due to uranium, thorium and potassium. Weight for weight, potassium contributes the least energy but it is more than a thousand times more abundant in the crust than the other two, and is an important source of heat. Radioactivity from the continental crust has been measured, and its magnitude accounts for about 70 per cent of the actual heat flow. The other sources must together make up the remaining 30 per cent, which is not much.

As for the ocean floor, where the heat flow is nearly as high as on the continents, even though radioactive heat production is much less, the possible origin of the heat flow is the object of much recent research. One important trend has already emerged: under the mid-oceanic ridges the flow is four or six times greater than elsewhere. This emphasises once more the uniqueness of such ridges and their importance, and the reality of their distinction from the continental mountain fold belts. Perhaps the high heat flow value is connected with volcanism, which is particularly active over the ridges.

Finally, what is going on deeper down in the mantle, and in the core? What is the temperature at various levels and at the centre? Many speculations have been made and the widely differing results which seem to be the most reasonable are shown in table 15.

Pressure inside the earth

The propagation of earthquake waves depends not only on the elastic properties of the propagating media but also on their densities. The results obtained are in good agreement with those taken from the boundary conditions imposed by the mean density of the earth, its polar flattening and the arrangement of its internal layers in order of increasing density. Combining the two, one finishes with the density values given in table 15 (pp. 78–9), which allow for just a little uncertainty (the margin is 20 to 25 per cent), except for the centre where one has a choice between twelve and eighteen.

Given the differences in density between the various layers within the earth, one can calculate the gravitational attraction which each layer experiences towards the centre of the earth. It must evidently be zero at the centre of the earth, but then moving away from the centre the force starts to increase, and the greater the distance away the stronger the force becomes because the quantity of matter between the point in question and the centre of the attracting mass is also greater. Then from a depth of about 3,000 kilometres right up to the surface the gravitational attraction remains roughly the same because the increased attraction due to lighter and heavier layers (the mantle and the crust) only just manages to compensate for the diminution caused by increasing distance from the centre.

Knowing the probable density of each layer, and the strength of the force of gravity, one can calculate the pressure at each level. A million atmospheres is reached about 2,200 kilometres down, and three million at about 4,700 kilometres. In the centre it must be around 3·6 million times atmospheric pressure. The uncertainty over the density distribution at depth hardly affects this value, for it is so enormous. Our best equipped laboratories can only manage pressures which are 36,000 times lower, that is 100,000 atmospheres, and even this is very difficult to achieve. What are the properties of matter under pressures of three or four million atmospheres? One cannot say, but certainly up to 100,000 they change considerably, and the arrangement of atoms within crystals is greatly modified. In addition, it could be that the atoms themselves change, or their distribution of electrons, or their dimensions.

Terrestrial magnetism

We have known since the Chinese discovered the lodestone that the earth causes a magnetised needle to set in a particular direction; in other words it has a magnetic field. Ninety-five per cent of this field arises from the interior of the earth, and since this probably consists largely of iron one would be tempted to imagine a huge magnet at the centre. A lodestone, however, loses its magnetic properties above a temperature of 580°C, which in the earth is reached at a depth of a mere fifteen or twenty kilometres.

The other possibility is that the earth is not a permanent magnet but an electromagnet, owing its magnetism to electrical currents in the earth. This cannot be the complete answer, for such currents would have to be impossibly large, more than 1,000 million amps. The magnetic field could be produced by electrical currents in a moving conductor, however, as in a dynamo. For this there must be a mobile medium which also conducts electricity, and in the interior of the earth there is only the core which appears to have the required mobility.

The earth's magnetic field may be resolved by calculation, although not without difficulty, into a principal component field acting approximately north-south, and six or seven secondary ones orientated in various directions. A map of this does not reflect anything of the geographical pattern of oceans and continents, in contrast with many other phenomena, and this is another reason for not attributing the source of the field to anything near the surface. The core seems to fit the conditions well.

3 Anatomy of the earth's crust

The earth has had a long history, 4,000 to 4,500 million years according to the current estimates, and during that time many changes have taken place on its surface and in its crust. The rocks of the crust have been heated, squeezed, folded and torn, and to unravel their history is a difficult task.

In reconstructing the history of the earth's form, and the forces which have moulded it, we shall need a guide. To facilitate our understanding of these forces, we may consider a piece of cloth lying on a table. If there were a tear in the centre of the cloth, we could imagine that someone had pulled at it strongly. If there were a fold, on the other hand, it could be that someone had pushed the cloth from two opposite directions. Things might be more complicated than this. Imagine there were two pieces of cloth on the table, approximately edge to edge, which could have arisen from a single cloth having been torn in two, or from two pieces having been brought close together. A fold could be produced by placing both one's hands flat on the table and pushing them towards one another, but it could also be caused by pushing them obliquely, and in nature analogous circumstances can arise. The cloth may be compared with the layer of rocks making up the continents.

Rocks classified by their mode of formation

Let us note first of all that by convention, soft and mobile materials like sand and clay are described as rock, as well as harder materials like marble and granite. There is no essential difference between, say, a sand and a sandstone, apart from their hardness. A sandstone is simply a sand which has become cemented into a hard material by various natural processes.

The first major group of rocks consists of clays, muds, sands and other materials which are formed from the disaggregation or alteration of other rocks by the external forces acting on the earth's crust, that is, the atmosphere, rivers, glaciers, the sea, and so on. Such rocks are termed *sedimentary* from the Latin word meaning *deposit*.

Other rocks, on the contrary, have once been molten and have crystallised by cooling, either on eruption as lavas at the surface of the earth, or by slower cooling within the earth's crust. These rocks are called *igneous*, from *ignis*, the Latin word for *fire*. Examples are basalt, the commonest kind of lava, which is a hard, fine-grained, black material, and granite, which owes its coarse texture to slow crystallisation far below the surface of the ground.

The third group arises from the transformation of members of the two preceding groups, mainly as a result of heat or pressure, whence their name of *metamorphic* rocks. Typical examples are the foliated mica-schist, gneiss which resembles a granite with the minerals distributed in bands, and marble, the metamorphosed equivalent of limestone. All rock types are susceptible to erosion and the products will form sedimentary rocks. These in their turn can become metamorphic rocks or gneisses under the influence of heat and pressure, possibly even culminating in melting and the production of new igneous rock types.

The known rates of some of these processes allow us to go into even greater detail. The rivers and glaciers carry between eight and fourteen cubic kilometres of solid matter into the oceans each year – let us call it eleven. If this has been happening on average at the same rate in the past as now, there will have been a layer of 30 millimetres average thickness deposited every million years allowing for compaction

Figure 30. An anticlinal fold seen from the air. On the left-hand side of the picture the beds of rock dip downwards towards the left and on the right-hand side they dip downwards towards the right. Folds such as this are clearly visible from the air in desert and semi-desert regions.

of the sediments, which would mean 100 kilometres since the formation of the earth. Now the earth's crust is only thirty kilometres thick under the continents and a good deal less under the oceans. One could thus suppose, in company with the American oceanographer Hamilton and the Norwegian geologist Barth, that a considerable portion of the deposits formed in the sea since the earth began have been transformed into metamorphic or igneous rocks. These have then been eroded once more, and the whole process started again. In this way the material of the earth's crust has not merely been transformed, but it has undergone a series of cyclic transformations, from sedimentary to metamorphic rock or vice versa, from metamorphic to igneous rock or vice versa, or from igneous to sedimentary rock.

The threefold division of rock types is as much of practical as of theoretical interest, for some minerals are found only in areas of sedimentary rocks, as coal, oil, and bauxite (the ore of aluminium), while others are mainly found in veins associated with igneous rocks, as, for example, the ores of lead and zinc. The distribution of the common rocks provides many other useful clues to the distribution of more valuable minerals.

Mountain fold belts, shields and platforms

To understand the structure of the land, its folds and other deformations, we need some means of measuring tilting and distortion, just as an architect needs to ascertain whether a surface is truly horizontal or not. The sedimentary rocks furnish us with just such a reference, since they were laid down originally in approximately horizontal layers. We see

Figure 31. Giant's Causeway, Ireland. The regular polygonal columns seen in the photograph are caused by the contraction of basalt lava flows during cooling. The Giant's Causeway is on the edge of the Antrim lava plateau.

Figure 32. The granite cliffs of Land's End
in Cornwall. Granite is the commonest type
of igneous rock to have crystallised
below the surface of the ground.

111

these layers in many places today, often tilted, deformed, and broken (figure 30). In metamorphic regions of sufficient thickness the beds of ancient sediments or lavas can be recognised, and these also are useful as reference planes. In this way the following different sorts of structural pattern may be recognised.

Mountain fold belts

Here the strata are inclined, the mean angle of inclination in different fold belts ranging from 10 to 72°. They have obviously been subjected to great lateral pressure, giving rise to both symmetrical and asymmetrical folds, overfolds and recumbent folds. In extreme cases a large folded mass of rock may be pushed over another for tens of kilometres; such a mass is called a *nappe* and it is said to have been *thrust* over the underlying rock.

The Rocky Mountains, the Alps, the Jura Mountains, and the Himalayas are all good examples of fold belts. They are clearly elongated in plan, being from three to ten times as long as they are broad. They are slightly curved rather than rectilinear. Their length and curvature are comparable to those of the island arcs, possibly indicating a relationship between these two structures. Almost a century ago the great Austrian geologist Suess said 'The growth of folds is hindered by obstacles which resist the transmission of horizontal pressure and they tend to develop along curved lines'. Persia, the Himalayas, the Rocky Mountains and the Andes are all good examples. The obstacle is usually a rigid block, visible in some instances, but concealed beneath younger and softer rocks in others.

Geographically the mountain fold belts are often situated

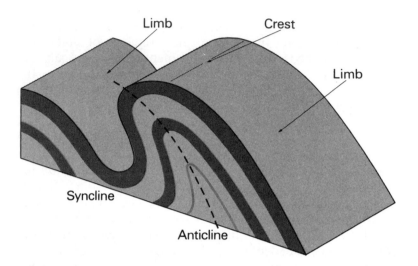

Figure 33. A slightly overturned fold. Such folds are common in many parts of the earth's crust, even though the rocks involved were originally deposited as more or less horizontal layers.

close to the border between continent and ocean. Like the island arcs, they form narrow belts around the continents. The two main examples are the circum-Pacific belt and the Mesogeic zone (Antilles – Mediterranean – Central Asia – Malaysia – Indonesia). Older fold belts sometimes occur further inland at the present day, but even in these cases their sedimentary strata yield a fossil marine fauna. The contact between sea and land has thus always favoured the occurrence of fold belts.

Their final characteristic, or temporally speaking their first, is that the sediments which make up most or all of them are very thick, as much as 10,000 metres in some cases and always thicker than those of surrounding undeformed areas. A remarkable feature of these sediments is that their fossils and their sedimentary structures show them to have been deposited in shallow water. We are not, therefore, considering a sea 10,000 metres deep which was filled with sediment, but a belt of shallow water deposition which subsided through 10,000 metres as the sediment accumulated. The belt of folding has thus been built up on the site of a former trough of sedimentation. Such a trough is described as a *geosyncline*.

Ancient massifs and shields

Other regions may display an area of rocks which have been folded, but subsequently dissected and lowered by the erosive action of the atmosphere, glaciers, rivers or the sea; such areas were once mountain chains, but are no longer. These areas are described as *ancient massifs*, or if large, *shields*, and good examples are the heart of Scandinavia and of Canada. As well as being eroded and levelled out, the ancient massifs have been uplifted. The rocks which are exposed at the surface today are not those which formed the upper part of the original mountains but are of a deeper origin; they are highly metamorphosed (the temperature and pressure being much greater at depth) and they are harder and more rigid.

Platforms

It may be that on top of areas of rocks which have been folded and planed, such as the shield areas, further sedimentary strata are laid down, and are preserved with relatively little folding or disturbance. This results in what is called a *platform*, and examples are to be found in Russia and the Sahara.

Figure 34. Grand Canyon, Arizona. The canyon has been formed by the downward erosion of the Colorado River. Horizontal beds of various sedimentary rocks are revealed in the cliffs along the sides of the canyon.

Distribution of folding in space and time

Whereas mountain fold belts are found along the continental margins, shields and platforms occur at the centre of the continents. One has the great shields of Canada, Scandinavia and northern Russia, lying to the north of the Mesogeic zone, and the Brazilian shield, the Afro-Arabian block, India and Australia to the south. Then, quite on its own at the South Pole, the Antarctic shield.

Around each of these shields are arranged fold belts of younger and younger age as one moves outwards. Any one fold system will not encircle the whole shield; far from it, they only take up a small part of the perimeter and always less than a half. Sometimes fragments of the ancient basement are caught up in the more recent fold belts and incorporated in their arcs.

In each region there have never been more than three or four epochs of folding, each one separated by long periods of quiet. The foldings have thus been intermittent, and the idea of cyclic events has arisen. Some of the major epochs of upheaval have been named, such as the Hercynian (350 to 230 million years ago) and the Alpine (beginning about 60 million years ago). In actual fact, as the dating of rocks is accomplished in greater detail one realises that there has been folding in all epochs, sometimes here, sometimes there, and the names which have been given to each period of folding are retained simply as convenient points of reference. Everything happens as though the earth's crust were constantly under a stress which is increasing in strength. To begin with, the crust resists the force, then when the limit of resistance is reached in a particular place the crust yields and folds in that area, and the stress is released for a while. Then it gradually builds up again until a new fold is produced either in the same place or elsewhere.

The spatial picture at which we have ultimately arrived – the old folded rocks in the centre with younger and younger folds towards the outside – calls for two comments. It is a true picture, but the present distribution of rocks on the earth's surface is very different from their original distribution. It is, moreover, certain that large areas of the continents have shrunk during the course of time because of having been folded. Many scientists even think that the different continents have been able to drift about, diverging or coming together.

Finally, no evidence of inclined strata or fold belts has yet been found on the ocean floors beyond the continental shelf and the extensions of island arcs. Here again there is a fundamental difference between the continents and the great oceanic floor.

Salt domes

The most abundant rocks have a specific gravity of 2·7, that is to say, they weigh two and a half to three times as much as water. Exceptions to this are rock salt and a few other minerals, which are lighter, with a specific gravity of 2·1, and are at the time more plastic and deformable. On account of these properties, whenever a layer of salt is overlain by a sufficient thickness of heavier rock it tends to rise and cut through the overlying strata. In this way very characteristic domes or columns of salt are formed, called *diapiric structures*. The pattern they form on a map is fundamentally different from that of folding. They are reminiscent of the small eruptions evident in some skin diseases or of the submarine volcanoes on the sea bottom.

Figure 35. The development of a geosyncline. In the first stage (*upper left*) a shallow sea separates two land masses several hundred kilometres apart; erosion of the mountainous areas on land (*lower left*) provides sediment which accumulates at the bottom of the sea and the sea bed sinks, partly due to the increased load and partly

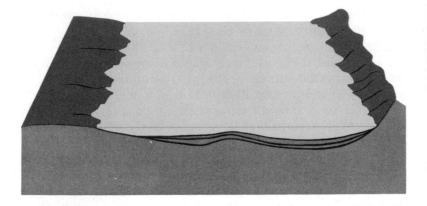

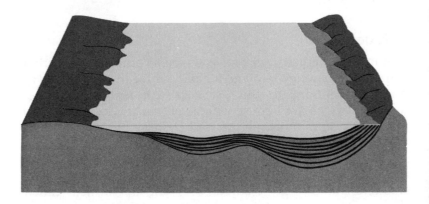

due to a downward pull from beneath; lateral compression causes emergence of some of the sediments as a row of islands (*upper right*); finally uplift and further lateral compression leave a mountain range of folded sedimentary rocks.

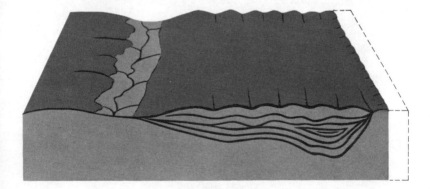

Figure 36. Arthur's Seat Edinburgh,
a volcanic plug surrounded by softer sedimentary
rocks on which the city is built.

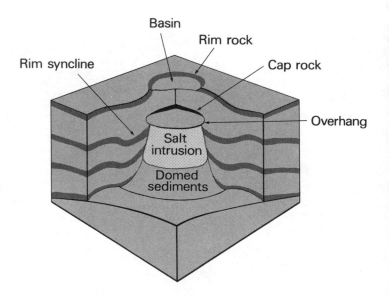

Figure 37. The form of a salt dome. Salt is so light and flows so easily that it tends to force its way into overlying sediments as a rounded plug. Many salt domes have been found below the North Sea.

The mechanisms of emplacement of plastic salt and viscous lava are somewhat analogous, perhaps due to the overlying weight while they are both still deep down.

Sometimes the rock salt actually breaks through to the surface and flows on the surface like a glacier, as in the mountains of Persia. In general, however, rock salt cannot often be seen at the surface because it is so readily dissolved by rainwater. Buried salt domes are more common, and are of great interest to geologists engaged in the search for oil because oil is often trapped in the domed strata above the salt. Many salt domes have been detected by geologists prospecting for oil in northern Germany and the North Sea. Another big field extends north-west of the Caspian Sea.

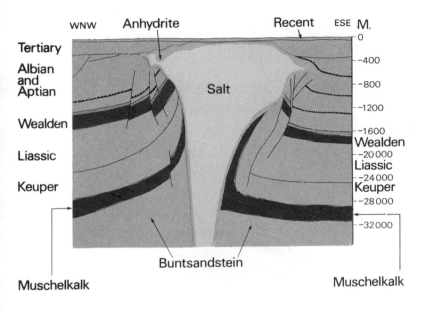

Figure 38. Section through a salt dome at Weinhausen – Eicklingen, Germany. The salt is dissolved by infiltrating water and few salt domes actually reach the surface, and most are known only from borings and from geophysical explorations for oil.

Great fractures of the earth's crust

The crustal rocks of the earth have not merely been deformed or folded under imposed stresses but have been fractured as well. Beds of rock have been displaced upwards, downwards or horizontally along more or less plane surfaces of fracture. Such breaks, with displacement, are known as faults. Faults may result in either an extension or a compression of the rocks. Faults are often found grouped in belts and also may form the boundaries of rift valleys, such as those of the Rhine Valley and East Africa, or horsts, such as those of Morvan or the Harz mountains. The strata near a fault are often slightly distorted, but dip only at a few degrees

Figure 39. Some major structural features of the earth's crust. Nappes and klippes are very severe folds which occur in many fold belts. Rift valleys are less common and occur outside fold belts in regions of tension. One of the greatest rift valleys extends from the Dead Sea to East Africa.

A nappe

A klippe

Figure 40. Overleaf the map shows the main
structural features of the earth's surface. (Based on several authors.)

125

considerably less than in a fold belt. Volcanoes sometimes occur along faults, as in Iceland where lava can be seen to have come from fissures in the ground. Faults may sometimes be seen at the surface in areas where earthquakes are common, but in time the unevenness due to a fault is reduced through erosion and eventually disappears completely.

Most faults are associated with fold belts. Some as thrust planes inclined gently at 10° to 25° indicate that a compression of the rocks has taken place; others with their planes inclined at an angle of 60° or so generally indicate the opposite, namely an extension, and are chiefly produced in the relatively quiet period following an episode of folding.

Other faults may be quite independent and are found outside the region of any recent folding. The most striking example on land is the huge rift valley which extends southwards from the Lebanon through Lake Tiberius, the Dead Sea and the Red Sea, and thence into East Africa taking in Lakes Rudolf, Albert, Kivu, Tanganyika and Nyasa. The width varies between 50 and 100 kilometres in Africa and reaches 300 kilometres in the Red Sea. Lake Tanganyika is up to 550 metres deep.

There are straight line escarpments on the floor of the ocean which also seem to be faults. The most remarkable are those discovered in the eastern Pacific since 1945, and extend approximately westwards from the coast of America across the Pacific Ocean. They consist mainly of four major ones which are between 2,400 and 5,300 kilometres in length and about 100 kilometres wide. With them are associated secondary escarpments and troughs. Islands and submarine mountains occur along the faults, although the latter are also present in the areas between the faults.

The discovery of the major faults described above

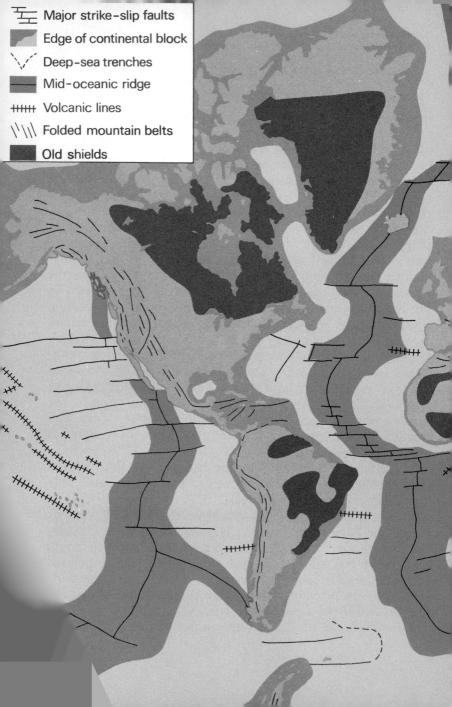

	Major strike-slip faults
	Edge of continental block
	Deep-sea trenches
	Mid-oceanic ridge
	Volcanic lines
	Folded mountain belts
	Old shields

encouraged some scientists to visualise a series of isolated blocks separated by a sort of world-wide network of criss-cross fractures. The diagrams which have been published are certainly very striking, but one ought not to be misled; the authors' talents for drawing and generalisation play a large part, as many of the supposed alignments are based on the extrapolation of faults which can actually be followed for only short distances. Mercator's projection in any case encourages a false impression of parallelism, because on a sphere such as the earth, lines are neither straight nor parallel, and meridians in particular are clearly convergent towards two common points at the poles.

There exists something intermediate between the supple deformation of a fold and the break due to a fault, called a *flexure*. As an illustration one may consider a sheet of paper, half of which is firmly pinned to a table and the other allowed to overhang the edge. If one brings a hand up sharply under the overlapping half it results in a step or displacement, as with a fault, but without any fracture. There is often a flexure between a region which is rising and another which is sinking. It is possible that there is frequently such a feature between the continents and the ocean floor, called a *continental flexure;* the east coast of Greenland is one example. Inland, between the mountains which border a continent and the plains of the interior, one could have an *intracontinental flexure*, but in many places this has developed into a fracture. The edge of the Atlas mountains against the Sahara is an example. Unfortunately the ancient Moroccan town of Agadir was built right on top of this feature, and in 1960 an earthquake destroyed it; the newly rebuilt town lies at a respectable distance. In general one would thus avoid constructing a dam or a tall building on the line of a fault.

Material rising from below

In addition to crustal deformations, and often more or less connected with them, there is the rise of *magma*, or molten rock, from deep down. Some of this reaches the surfaces as lava, while in other instances the material stops short of the surface forming volcanic *plugs*, *veins* and *dykes*, and more deep-seated bodies, called *plutons*.

Volcanic plugs have a mean diameter of two hundred metres, but may be as wide as 1,500. The rock forming the plug is often more resistant than its surroundings, and when exposed by erosion it forms a projecting peg which dominates the landscape (figure 41).

If the two sides of a more or less planar fissure are forced apart and the intervening space is occupied by magma, we have a dyke. Sometimes a fissure is occupied by a variety of minerals deposited from hot solutions or vapours, in which latter case it may often be a valuable source of metallic ores, such as those of zinc, lead, silver and gold. The largest dyke known is the Great Dyke of Rhodesia, 480 kilometres long by eleven kilometres wide. Normally, however, dykes are of more modest dimensions.

Mineral veins may be seen and admired in rocks exposed along the sea shore or in mountains, cutting through the surrounding rock and varying in width from a few millimetres to several metres. They are often composed mostly of white quartz, a sort of silica, or sometimes of calcite.

It was dykes that fed the great basalt flows of the Deccan plateau in India, and of Iceland. Instead of cutting throu the surrounding rocks, the magmas have sometimes b insinuated between beds of other rock, forming *sills*. T horizontal bands of black basalt stand out strikingly

Figure 41. Devil's Tower, Wyoming –
a mass of hard igneous rock which
stands out from the softer sedimentary
rocks which surround it.

131

rocks above and beneath are light in colour. Fine examples
are to be found in the Royal Society mountains of Antarctica,
one of whose sills is six hundred metres high and stretches for
twenty kilometres. How can the basalt have forced its way
between the layers of sandstone, raising them so regularly
from beneath? It was evidently accomplished under pressure,
and without any sudden jerk, for otherwise the basalt would
have broken through to the surface and formed a volcano.

In a given region dykes and sills often occur after a period
of folding. Subsequently the surrounding rocks may be
eroded and often the dykes are left projecting like walls, in
which case their name is particularly apt.

Whereas volcanic plugs are cylindrical and dykes or veins
are planar, plutons extend in a more massive three dimensional form, and are often justly called massifs. Solidifying at
depth, beneath 1 to 10 kilometres of overlying rock, they
never reach the surface of the ground. The smallest examples
are of the order of a few kilometres in length; one of the
largest, the granite massif forming the coastal mountain
range in British Columbia, stretches for 1,700 kilometres
with a width varying from 30 to 230 kilometres.

Massifs are most commonly formed of granite, and are
found especially in the heart or deepest portion of fold belts,
whether ancient or recent. They also make up an important
part of old shield areas which have been strongly scoured by
erosion; a quarter of the Finnish shield is constituted o
granitic rocks. Massifs are slightly elongated in for
following the long axis of the mountain range in which t
are found. According to a few geologists they were for
that is to say, their constituents rose or at least consol
during the period of folding, but the majority of ge
would say it was mostly after the folding.

In a similar way to dykes, veins and volcanic vents, the plutons usually cut clearly across other rocks. They often have an irregular contact with the overlying rock and their roof, to use a mining term, is hummocky in form. Portions of the roof frequently appear to have collapsed into the magma before it consolidated. A granite emplaced during a period of folding follows more closely the form of the surrounding rocks, and good examples of this are found in Scandinavia.

Sometimes, particularly in a shield area, a gradual transition may be seen between an igneous rock and the metamorphic rocks into which it appears to have been intruded. This leads one to believe that not all the matter in a massif has risen from the depths but that to a considerable degree it has been produced by the reconstitution of rock which was already there, and that some transformation has taken place of which the igneous rock represents the end-product. There is no doubt that the pressures and temperatures necessary for such transformations may well have existed at the centre of fold belts, deep in their roots. This may provide an explanation for the fact that plutons almost never distort the rocks which enclose them when the two come into contact; the latter have somehow been consumed and digested on the spot by the material of the pluton. Whether granite and the other igneous rocks have existed in a liquid or viscous state or whether they have been able to form by simple recrystallation while still solid is an open question, but of course re is no doubt about the volcanic rocks.

ng after the formation of a massif it is possible that it, or of it, may be re-disturbed during more recent folding. tras granite in the Carpathian mountains is a good In this case the ancient granite is seen to bear the marks of more recent fracturing and shearing, evidence of the mechanical stresses to which it has subsequently been subjected.

Recent movements of the earth's crust

All the features and phenomena which we have just been considering were reconstructed from evidence which, although certain, was often indirect: the inclination of strata, displacements, and shortenings. There are, however, many direct indications of earth movements, including of course earthquakes. The most violent movements are accompanied by fracturing of the crust along faults; record displacements of the ground on one side relative to the other are ten metres vertically in Assam in 1897 and seven metres horizontally in San Francisco in 1906.

Faults can move slowly along sharply defined lines. Engineers on the Buena Vista Oilfield in California could hardly have been more surprised in 1933, when their roughly buried pipelines started to rise in contortions out of the soil. That the ground was shortening was confirmed in boreholes where the casings showed a deformation which could be aligned from one bore to the next, exactly along the plane of a fault. It was measured to be moving at a rate of thirty-eight millimetres a year.

Most slow movements, however, cause deformation without fracture. The solid ground is known in some cases to rising. Travelling out of Oslo along a road bound for mountains, one may find mussel shells far up in the fo for a few thousand years ago this was at the edge of and since then the land has been uplifted. This is over the entire Scandinavian peninsula. Huge

Figure 42. Map of Scandinavia showing the amount of uplift (in metres) which has taken place since glacial times, as shown by raised beaches, marine shells etc. The uplift is due to the removal of the ice-cap which once covered Scandinavia (see figure 43). Isolines according to Finnish and Scandinavian specialists.

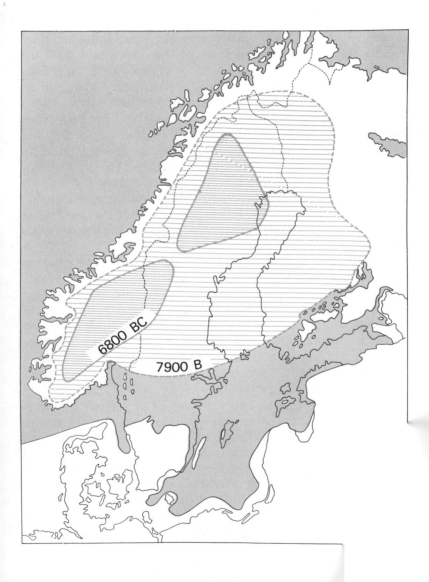

Figure 43. Former extent of the Scandinavian ice-cap. By 6,000 BC the ice-cap had largely melted, but post-glacial uplift has continued to the present day.

covered it up to 10,000 years before our time, and when they melted the sea flooded back on to the coasts and built up beaches, where the remains of a characteristic shellfish, *Yoldia*, were deposited. The land subsequently rose, and at the present day Yoldia shells are found in beach deposits up to 275 metres above actual sea level. The land probably rose because the weight of ice had been removed, a delayed reaction which is still happening today, as demonstrated by tidal records and geodetic measurements. In the Gulf of Bothnia the rate of uplift is one metre a century, and in Stockholm thirty centimetres. This has a considerable practical effect in that the Baltic ports are faced with a falling water level, and must either dredge their harbours or be replaced as the sea retreats.

Estonia and Latvia are actually rising at a rate between 3,000 and 6,000 millimetres in a thousand years, Moldavia at 8,000 millimetres and Scotland at 4,000 millimetres. These uplifts are partly compensated for by an overall rise in world sea-level which has been about one to two millimetres a year since the beginning of this century. Other evidence is provided by the littoral deposits of two million years ago which are found today 1,283 metres above sea level on the island of Timor in Indonesia.

Ancient lake deposits may also serve as reference points. Lake Bonneville in Utah once extended over 90,000 square kilometres, reaching a depth of more than three hundred metres. As the climate became drier, the water level sank and are left today with the modest testimony of the Great Salt e. In this area the land, relieved of a load of ten million of water, has risen. Around the borders of the lake and ral islands in its centre, measuring the actual positions r beaches, one sees that there is a relative difference in altitude which clearly demonstrates that the land once under the lake has been deformed. The greatest difference in altitude of the supposedly contemporaneous beach deposits reaches a maximum of sixty-four metres in the centre of the old lake, exactly where the burden of water was greatest, and the difference decreases in remarkable correspondence with decreasing depth of water in the original lake. In other regions the land is sinking, the Netherlands for instance at a rate between 1·0 and 2·6 metres in a thousand years, and also the Po delta.

In some regions such as those of volcanic activity the land may alternately rise and sink. On the edge of the Bay of Naples there is an ancient Roman market incorrectly known as the Temple of Serapis. It was obviously constructed above the level of the nearby sea, but today one can see holes bored in the three remaining stone columns by marine bivalves in that part between 3·65 and 6·35 metres above the base, some containing the shells of these molluscs and other marine organisms. There must thus have been a downward movement of at least six or seven metres sometime in the Middle Ages. Around 1530, however, the ground began to rise again, and in 1538 the columns were once more entirely exposed to air, just at the time when the volcano Monte Nuovo arose as a result of a local eruption. Since 1800 the columns have undergone minor movements of one or two centimetres a year. Sometimes their base is dry and sometimes submerged, as in 1944.

As well as vertical movement, very precise topographi measurements have revealed slow horizontal moveme especially in the Soviet Union and the United States great San Andreas fault in California shows a relative ment of as much as five centimetres a year.

We now turn to the ocean floor. The slow movements which have been detected there are mainly in a downward direction. Around the Aleutian Islands information gathered by the US Navy has revealed a very clear relief with valleys and tributaries, obviously formed sub-aerially and submerged now at a depth of 360 metres. Even allowing for a rise in sea level, there remain no less than 260 metres which are inexplicable by anything other than a sinking of the land. Around the island of Ceram in Indonesia the tops of coral reefs are found at a depth of 1,500 metres, although we know that they must have been formed at sea level because coral polyps can live only in very shallow water.

Boreholes on the atolls of Bikini and Eniwetok in the middle of the Pacific provide further basic information. Coral reefs from twenty-five million years ago, as shown by their fossil content, are found seven hundred metres beneath the present sea level. The deeper Eniwetok boring produced fossils sixty million years old at twelve hundred metres; this gives the mean rate of sinking as twenty millimetres in a thousand years.

We are provided with striking evidence by the more or less conical submarine mountains which we know to be the remnants of ancient volcanoes. About six or seven per cent of them, called *guyots*, appear from the published charts to have a truncated flat top, and rounded pebbles have been dredged from these surfaces, indicating that they have been eroded by ʾaves and were thus once at sea level. In addition, shells of ⁺inct shallow-water animals have been found on the guyots. ⁻eneral, the lower the top surface of the guyot is below ⁻vel, in other words the greater its total lowering, the ⁻re the fossils found upon it. The most ancient go back ⁻ed million years, and include the Hess and Cap

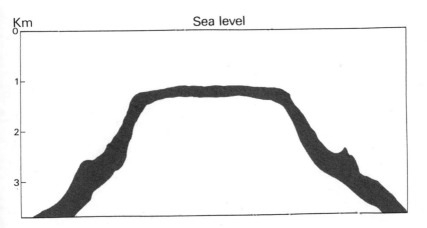

Figure 44. Record made by an echo-sounder across a typical guyot in the western Pacific. The flat-topped sea-mount is several kilometres across.

Johnson guyots whose tops are about eighteen hundred metres under water. Their mean rate of sinking must have been eighteen millimetres in a thousand years, which compares closely with that of Eniwetok (twenty millimetres).

From this we can conclude that certain parts of the ocean floor have a tendency to sink. Perhaps the effect is due to the overburden of mud, which is deposited there at a comparable rate: 30 millimetres in a thousand years. It is probable however, that the rate of sinking, like that of sediment deposition varies from one region of the ocean to the next.

To summarise, the rocky crust of the earth under both air and sea is moving and changing all the time. The rate of movement is small on the scale of human history, but perhaps this is just as well for the forces involved are immense, and far beyond our power to control.

4 Chemical constitution of the earth

The universe is composed of particles, including *protons*, *neutrons* and *electrons*, which on grouping together form the atoms of the different elements, such as lead or gold. These particles differ from one another in their mass and in their electrical charge. The proton and the neutron have the same mass, which is considerably more than that of the electron. The electron bears a single negative electrical charge, and the proton bears an equivalent positive charge, while the neutron is electrically neutral. In all atoms the number of electrons is the same as the number of protons, so that the overall electrical charge is zero and the atom as a whole is neutral. The simplest atom is that of hydrogen, with one proton and one electron. The next simplest is that of helium, with two protons, two electrons and two neutrons. In this way the atoms of the 103 known elements are built up, of which about ninety occur in nature.

Each element is characterised by the number of protons (and electrons) contained by its atoms, and this number is called the *atomic number* of the element. The number of neutrons in the atom may vary, however, and the total number of protons and neutrons, upon which the total mass of the atom depends, is called the *atomic weight*. Atoms with the same atomic number and different atomic weight are described as *isotopes* of an element.

By analysing the spectrum of light from the sun and stars it is possible to know the relative distribution of various elements throughout parts of the universe. Chemical analyses meteorites which have fallen on to the earth from space contribute to this knowledge. The result is not what one expect (table 18). By far the most abundant of ts in the universe is that with the smallest and simplest ydrogen. The next most abundant element is helium,

Table 18 Abundance of the principal elements
expressed as the numbers of atoms relative to a basis of 10,000 atoms of silicon, after Brian Mason

Element	Z	A	The Universe	Earth	Earth's crust
Hydrogen	1	1	400,000,000		1,430
Helium	2	4	31,000,000	X	X
Oxygen	8	16	215,000	39,400	29,400
Neon	10	20	86,000	X	X
Nitrogen	7	14	66,000	X	3
Carbon	6	12	35,000		27
Silicon	14	28	10,000	10,000	10,000
Magnesium	12	24	9,100	15,000	865
Iron	26	56	6,000	14,000	900
Sulphur	16	32	3,750	1,900	17
Argon	18	40	1,500	X	X
Aluminium	13	27	950	340	3,400
Calcium	20	40	490	336	910
Sodium	11	23	440	135	1,240
Nickel	28	59	270	21	1
Phosphorus	15	31	100		39
Chlorine	17	35	90	4	6
Chromium	24	52	78	35	4
Manganese	25	55	69		18
Potassium	19	39	32	40	670
Titanium	22	48	24	18	94
Cobalt	27	59	18		X
Fluorine	9	19	16		38

The blank spaces indicate an absence of data

Z = atomic number

A = atomic weight (approx.)

X denotes less than one atom

and in general, granting a few irregularities in detail. The more electrons an atom has and the greater its complexity, the rarer is its occurrence in the universe. As a guide in comparing two elements, if the number of electrons is doubled, the abundance is divided by about one hundred. It seems that in distributing the immense number of elementary particles, nature has used mainly the simplest and most straightforward combinations. Furthermore, we find that apart from hydrogen, elements with an even number of electrons are more abundant than those with an odd number.

The chemical elements in the earth

Does the earth constitute a representative sample of the universe, or is its composition unusual in any way? In order to answer this, we may get some idea from analyses of the atmosphere, the ocean and of the crustal rocks. In the case of the inaccessible deeper layers, we are reduced to making comparisons, of which the most valuable would seem to be with meteorites. We then find that the earth's composition (table 18) is very different from that of the universe, and it is easy to understand why. Whereas the universe is composed mainly of gaseous matter, a large proportion of the gases have escaped from the earth, probably because its force of attraction was not sufficient to retain them, especially in its early history when they were hotter and more mobile, and when the earth was spinning at greater speed, thus generating a greater centrifugal force. In this way the relative paucity of hydrogen, helium, neon, nitrogen (except in the atmosphere), argon and chlorine are accounted for. Carbon, which is extremely rare, could have escaped in gaseous form as one of oxides, or as methane, acetylene or numerous other gases.

Table 19 Composition of successive layers of the earth
in per cent of atoms

	Nitrogen	Hydrogen	Carbon	Oxygen	Silicon	Metals
Atmosphere	76	2	X	21	0	0
Terrestrial ice	X	66	X	33	0	X
Living organisms	1	60	8	30	X	X
Oceans	X	66	X	33	X	X
Crust	X	3	X	60	20	16
Mantle (estimated)	X	X	X	54	14	29
Core (estimated)	X	X	X	X	X	100
Whole earth	X	X	X	48	12	37

X = less than 1 per cent

Oxygen alone among the gaseous elements is relatively abundant on earth, and this is probably because of its remarkable ability to form stable compounds with other elements, making molecules which are too heavy to escape.

If the earth has lost most of its gases, by the same token it has been relatively enriched in the heavier elements, especially metals such as iron and magnesium. Silicon, a close relative of carbon, is more abundant than the latter probably because silicon combines with oxygen to form a solid, whereas carbon forms gaseous oxides which have been more easily lost from the atmosphere.

Table 20 Succession of layers which constitute the earth

	Mean thickness in metres	Mean relative density	Mass in million million metric tons
Atmosphere	300,000	0 to 0·0013	5,200
Terrestrial ice	60	0·91	28,000
Living organisms	0·1	1·00	20
Oceans	2,630	1·04	1,390,000
Crust	17,000	2·80	24,000,000
Mantle	2,883,000	4·50	4,075,000,000
Core	3,469,000	10·71	1,876,000,000
Whole earth	6,371,000	5·52	5,976,000,000

Layers of the earth

The earth is formed of successive shells, whose densities and thicknesses are given in table 20. Some of the shells are accessible for observation: the atmosphere, the ice-sheets, the assemblage of living organisms, the oceans, and the rocky crust which extends under the oceans. Deeper down are the mantle and the core, revealed by the propagation of earthquake waves. Some rocks come from the depths of the earth and give us a glimpse of the mantle, or at any rate, its upper part. We shall nevertheless be more sure of ourselves when we can actually investigate the mantle directly, and indeed a project was started a few years ago to drill a borehole down to reach the mantle (the Mohole project), but unfortunately the enterprise proved too expensive. As for the earth's core, the day when man will be able to reach this appears to be far distant!

From one known layer to the next, the density increases, and it seems as though matter within the earth has been arranged in order of its density. This would indicate that at some stage, perhaps at its origin, the earth was fluid, and as we have seen, the relative flattening of the poles gives a similar indication. There is every reason to suppose that the increasing density continues through the inaccessible layers of the lower mantle and core.

It may be seen from the densities and thicknesses in table 20 that the greatest part of the earth's mass is taken up by the mantle, followed by the core. The crust which means so much to us in terms of mineral wealth and so on forms merely four thousandths of the whole! The remainder, including the oceans, is even more negligible on the scale of the entire planet.

The uncertainty which hovers over the nature of the core and the mantle does not detract from the validity of the data on their mass and probable density given in table 20. These have in fact been calculated from the total mass of the earth, which is well established from astronomical evidence, and from the $1/298 \cdot 3$ flattening. Further investigations have shown that other density determinations of the mantle and the core lead to very similar results.

Composition of the sea

Hydrogen has a particular affinity for oxygen, and in the earth most of the hydrogen has combined with oxygen to form water. Very little hydrogen is left in the rocks and the atmosphere. Being lighter than the rocks, the water has concentrated above them as the oceans and seas.

Great interest lies in those substances which are dissolved in water (table 21). While sea water is especially rich in sodium and chlorine, the two well known components of common salt, the water poured into the sea by rivers is quite different: the predominant ions are those of carbonate and

Table 21 Mean composition of the solid matter dissolved in sea and river water
expressed as grams per cent

		River	Sea
Carbonate	CO_3^{--}	35·15	0·41
Sulphate	SO_4^{--}	12·14	7·68
Chloride	Cl^-	5·68	55·04
Nitrate	NO_3^-	0·90	X
Calcium	Ca^{++}	20·39	1·15
Magnesium	Mg^{++}	3·41	3·69
Sodium	Na^+	5·79	30·62
Potassium	K^+	2·12	1·10
Iron and Aluminium	$(Fe,Al)_2O_3$	2·75	X
Silica	SiO_2	11·67	X
Sundry others	Sr^{++}, Br^-, HBO_3		X 0·31
Total		100·00	100·00

X = less than 0·01 per cent

calcium, the two constituents of calcite, and silica is also relatively much more abundant than it is in sea water.

There is a continuous circulation of water on the surface of the earth. The rivers provide the sea with its water and its dissolved substances. The sea water evaporates and forms clouds, these precipitate as rain, and the rain water runs into rivers. Thus an equilibrium is established by means of this more or less closed cycle of water movement. This is not, however, the case for the dissolved matter, which does not evaporate, and must gradually accumulate in the sea. What does this mean, given that there is no leakage and no other means of entry? In the sea the uptake of dissolved salts by living organisms, or the precipitation of soluble matter when its concentration becomes too high, or the two together, must cause the preferential removal of certain elements, and thus alter the composition of the sea water. It is no accident that the substances dissolved in excess in river water and then transported to the sea, calcite and silica, are just those which living organisms use to form their internal and external skeletons. Corals and molluscs have skeletons and shells of calcium carbonate, and diatoms and Radiolaria have skeletons of silica, while sponges may have skeletons of either material.

Composition of the atmosphere

The main constituents of the atmosphere are nitrogen and oxygen, both of which are relatively light as one would expect in the outermost gaseous layer. A little hydrogen exists in combination as water vapour.

The quantity of free oxygen is astonishing for an element so ready to combine with others. The crustal rocks contain iron, manganese and sulphur minerals which are not saturated with oxygen and in such quantity that a layer of crust two hundred metres thick would be sufficient to fix all the free oxygen in the atmosphere. Since, despite the stirring action of erosion, deposition and other phenomena over three or four thousand million years, free oxygen is still present in the atmosphere, there must be some special reason for it. We shall see that the key to the mystery is the activity of living organisms, and this may also partly explain the surprising abundance of free nitrogen.

Atoms of living organisms

In terms of the mass of the earth, animals and man weigh an insignificant amount and they contain 250 times less matter than the atmosphere. Somewhat paradoxically, the

Table 22 Comparison of chemical compositions of the principal rocks of the crust and mantle

	Silicon (SiO_2)	Aluminium (Al_2O_3)	Iron ($FeO+Fe_2O_3$)
Granite	70	14	3
Andesite, Diorite	57	17	8
Continental crust (Goldschmidt)	61	16	7
Continental and oceanic crust (Poldervaart)	55	15	9
Basalt, gabbro	49	16	11
Peridotite (upper mantle ?)	42	3	8
Stony meteorites	42	3	19

composition of organisms resembles water; over sixty per cent of their mass is water, and this may reach ninety per cent in some organisms. It is only a short step from this to the idea that living organisms first evolved in water. Two other atoms are also characteristic of living organisms: nitrogen and carbon. Nitrogen is abundant only among organisms, and in the atmosphere; elsewhere it is found only in infinitesimal quantities. Carbon is also greatly concentrated in living organisms.

The extreme diversity of animals and plants is largely due to the exceptional ability of carbon to form varied combinations. This follows from the fact that one carbon atom will join simultaneously to four other atoms, either identical or different. It is the smallest, lightest and most mobile of all

By convention the chemical compositions of rocks are expressed as weight percentages of the oxides of the elements

Magnesium (MgO)	Calcium (CaO)	Sodium (Na$_2$O)	Potassium (K$_2$O)	Density
1	2	4	4	2·6
4	7	3	2	2·7
3	3	2	4	2·7
5	9	3	2	2·8
7	10	3	1	2·9
41	2	X	X	3·3
24	3	1	X	3·4

X = less than 0·5 per cent

the atoms which are quadrivalent. One may easily understand that living organisms have to some extent monopolised the carbon, and left the neighbouring envelope of atmosphere and crust relatively impoverished in this element.

The rocks of the earth's crust

There are numerous chemical analyses at one's disposal, but the difficulty in estimating the composition of the crust lies in appreciating the quantitative distribution of the multitude of different rock types which geologists have been able to distinguish. The American geochemists Clarke and Washington have worked out that the upper 16 kilometres of the crust consist of 95 per cent igneous rocks, 4 per cent

Figure 45. Gabbro with a small amount of quartz.

shale, 0·75 per cent sandstone and 0·25 per cent limestone. Other scientists have arrived at very similar answers and the most probable compositions of the main rock types and the crust are given in table 22.

In terms of abundance, two kinds of igneous rock are much more important in the crust than all the others, the basalts and the granites. In both types of rock the principal constituents are silicon and oxygen with a smaller amount of aluminium, but they differ substantially in their minor constituents. Calcium, iron and magnesium are important in the basalts, while sodium and potassium are important in the granites.

The sedimentary rocks are more varied in their composition because of their complex histories, while metamorphic rocks can correspond in composition to any of the igneous or sedimentary types from which they have been produced. However, as most authorities are agreed that at some time in its history the earth has passed through a molten stage, we may regard all the sedimentary and metamorphic rocks as products of the igneous rocks from which they were first derived.

Let us consider the distribution of these various rock types in the earth's crust. First of all we have abundant evidence that the ocean floor is made of basalt, underneath a layer of mud, and the great majority of volcanic oceanic islands are also made of basalt. It is basalt which is dredged up from the submarine oceanic ridges. One finds it on the crest of the mid-Atlantic ridge, together with gabbro and a small quantity of peridotite or related rock. *Gabbro* has the same chemical composition as basalt but its constituent crystals are bigger, like those of granite, and like the latter it solidified at depth. This all demonstrates that the oceanic crust is basaltic, and

that near the submarine ridges a tendency for crustal rupture which we have already noted allows matter of even deeper origin to emerge.

If we move from the open ocean towards the continents we encounter another type of volcanic rock – *andesite*. This is most striking around the Pacific where the '*andesite line*' separates the exclusively basaltic central area from a border zone of andesitic and other rocks which are richer in silica, lighter, and thus of a continental type. In the east this line runs closely along the American coast; in the west it runs outside the most seaward of the island arcs. Thus most of New Zealand, the archipelagoes of Fiji and the Solomon Islands, New Guinea, which is part of the Australian continental block, and also the island arcs of the Marianas, Bonin, Japan, the Kuriles and the Aleutians, are left in the continental domain. The chemical resemblance between the island arcs and the continental crust is one more reason for considering them as mountain chains in the process of formation, and the continents as being augmented along their coast by successive island arcs which have accreted there in the course of time.

The mass or at least the upper part of the continental crust is most closely comparable to andesite, granite, or a mixture of the two, considering that its mean chemical composition, which as shown in table 22, is distinct from that of basalt. One fact, however, requires some consideration. There are many occurrences of basalt on the continents and some are so imposing and extensive, for instance in Ethiopia, India and Siberia, that one is forced to think of molten or potentially fusible masses of truly basaltic chemical composition deep down even under the continents. Are these isolated patches or are they connected to the basaltic crust under the oceans? In the former case one must imagine that the continental crust has a second, deeper, denser layer of basalt under the well established uppermost layer of granitic or andesitic material. Putting the question another way, does the basaltic material under the continents consist of separate islands, or a discontinuous layer, or a thin shell, or ultimately of a continuous sheet? Most recent works suggest a continuous sheet. Before this interesting question is settled one ought to recognise that the continental basalts are not precisely equivalent to the oceanic ones. In particular the latter are more persistently rich in *olivine*, a beautiful yellowish green transparent mineral sometimes used as gemstone under the name of *peridot*.

In addition to their different distributions, the behaviour

of basalt and granite magmas is also different. The former tends to erupt from volcanoes and only solidify on or not far beneath the ground surface, whereas granitic magma crystallises more often at depth, giving the great enclosed massifs of the interior which we have called plutons.

The duality of the basalts and granites is too marked for us to ignore its possible causes. Why in particular is one so dominant in the centres of the oceans, and the other confined to the crust of the continents? Many scientists envisage the outer part of the earth, early in its history, as a continuous layer of molten basalt, to the top of which, as the whole began to solidify, floated a scum of light granitic material, while heavier material of peridotitic composition tended to sink to the bottom. The patches of granitic scum may be thought of as the beginnings of the continents. This is, however, only a hypothesis, perhaps useful when thinking of the origin of things, but what has happened since? Our lands and seas have had a very long history. The material worn from the continents by erosion is carried by the rivers towards the sea, and the sand particles which are the most rich in silica settle especially close to the coast; these are later incorporated in the coastal fold belts and contribute to the formation of continental granites whose high silica content is thus explained. Later, when these granites are exposed to the air at the ground surface they are re-eroded and the whole story starts again. This has happened not once, but again and again; in the earth's history there have been many cycles of erosion and sedimentation.

In spite of the great variety of igneous, metamorphic and sedimentary rocks, two elements are more abundant than the rest: oxygen and silicon. It may seem curious at first sight that the most abundant element in the solid crust is oxygen,

Figures 46 and 47. Left hornblende
andesite (*top*) and pyroxene andesite (*bottom*).
Below worn crystals of olivine, the major
mineral of peridotite.

which in its free state is a gas. It is predominant in both mass (45 to 47 per cent) and volume (94 per cent) throughout all rocks. The latter is due to the size of the oxygen atom, greater than that of any other common element.

Silicon is the next abundant in mass (26 to 27 per cent) but occupies less than one per cent in volume. It nevertheless plays a fundamental part in the formation of rocks and minerals, analogous to that of the carbon atom in living organisms, because like carbon it is quadrivalent and will link four other atoms simultaneously. The basic three dimensional building unit of the silicate minerals is formed by a silicon atom linked to each of four oxygen atoms, which

Table 23 Probable mean composition of the earth's crust

by weight in parts per million (p.p.m.), or grams per metric ton, and by volume

	By weight		By volume
	Continental crust (after Washington & Mason)	Continental and oceanic crusts (after Poldervaart)	Continental crust (after Washington & Mason)
Oxygen	466,000	449,000	937,700
Silicon	277,000	258,000	8,600
Aluminium	81,300	81,000	4,700
Iron	50,000	65,000	4,300
Calcium	36,300	63,000	10,300
Sodium	28,300	22,000	13,200
Potassium	25,900	16,000	18,300
Magnesium	20,900	31,000	2,900

from 9,000 to 1,000 p.p.m.	Titanium Hydrogen Phosphorus Magnganese
from 700 to 110 p.p.m.	Fluorine Sulphur Strontium Barium Carbon Chlorine Chromium Zirconium Rubidium Vanadium
from 80 to 10 p.p.m.	Nickel Zinc Nitrogen Cerium Copper Yttrium Lithium Neodymium Niobium Cobalt Lanthanum Lead Gallium Thorium
from 7 to 1 p.p.m.	Samarium Gadolinium Hafnium Dysprosium Tin Boron Ytterbium Erbium Bromine Germanium Beryllium Arsenic Uranium Tantalum Tungsten Molybdenum Caesium Holmium Europium
from 0·9 to 0·1 p.p.m.	Terbium Lutecium Mercury Iodine Antimony Bismuth Thulium Cadmium Silver Indium
from 0·09 to 0·01 p.p.m.	Selenium Argon Palladium
from 0·005 to 0·001 p.p.m.	Platinum Gold Helium Tellurium Rhodium Rhenium Iridium Osmium Ruthenium
less than 0·001 p.p.m.	Neon Krypton Xenon Radium

in turn are linked to other atoms such as another silicon, or sometimes another metal, and so on.

Among the metallic elements in rocks, aluminium is the most common, and from table 18 it may be seen to be ten times more common in the crust than throughout the earth as a whole. For potassium this factor is as much as sixteen. Still other atoms are represented in the crust, as shown in table 23 but surprisingly enough many of the useful ones are really quite rare. We shall be returning to this matter later.

The mantle

Unlike the crust, which can be seen and examined at the surface, the mantle has never been seen, and we only have indirect evidence of its composition. The main evidence, as we have seen, is that of the velocity of earthquake waves. The number of possible materials from which the mantle can be made are limited to those which transmit earthquake waves with the same velocity as the mantle. There are two principle candidates, peridotite and eclogite, both of which occur in the crust. Peridotite is a heavy dark-green igneous rock. Although not as common as basalt or granite, it occurs in many parts of the world, and forms, for example, much of the picturesque peninsula of the Lizard in Cornwall. Eclogite is much rarer, and is a distinctive rock containing large crystals of red garnet and a green mineral called jadeite. Peridotite is mainly composed of magnesium, oxygen and silicon, while eclogite is chemically very similar to basalt.

Eclogite has been proposed as a possible mantle material because it has the same chemical composition as basalt, but it differs from it in being much more dense. We know, of course, that the mantle is much denser than the crust, and in

Figure 48. Coastal erosion at Mayon cliffs, near Land's End, Cornwall. Huge blocks of granite have fallen from the cliffs and will be rounded and eventually broken down by wave-action.

Figure 49. Peridotite. Usually it looks darker.

fact it has been shown experimentally that at very high pressures basalt is converted into eclogite. Seismologists tell us that the earthquakes in volcanic regions often occur deep in the mantle, and some scientists think that this is where the lava comes from. If the mantle were made of eclogite, then the world-wide eruption of basalt lavas is easily explained, for if an eclogite were melted a basalt lava would be produced.

The difficulty with this hypothesis is that if the mantle is chemically similar to the basalts of the crust, and the boundary between the crust and mantle is a physical change,

Figure 50. Eclogite from Laxpana, near Kandy,
Ceylon. Eclogite is a rare rock type at the surface
of the earth, but some geologists think the mantle
of the earth may be made of eclogite.

due to an increase in pressure, and not a chemical change,
then it should happen everywhere at the same depth where
the pressure is the same. As we have seen, the boundary is
not always at the same depth; the crust is much thicker under
the continents than under the oceans. Nevertheless the idea
of a physical change is worth remembering because it may
explain some of the minor discontinuities which are known to
occur within the mantle.

The alternative possibility, peridotite, also has the right
physical properties, although some scientists think that it

does not explain the origin of basalt lavas as well as eclogite. However, we often find that the lavas erupted at the surface contain inclusions of solid material brought up from the depths, and these are more often peridotite than eclogite.

There is another type of rock which is thought to be derived from the mantle and which has always attracted interest because of its scarcity – *kimberlite*, the material of the South African diamond pipes. Kimberlite occurs in only a few areas of the world, and everywhere it has been excavated for its precious content of diamond. Diamond is really nothing more than a form of carbon, and it has been made artificially in recent years by the simple method of subjecting carbon to very high pressures. It can be calculated that the pressures required to form diamond are so great that it could only be produced within the earth at depths of 100 kilometres or more, and the diamond which is found in kimberlite pipes must therefore have come up from the mantle. Again we come back to peridotite, for kimberlite is very similar to it, and contains a high proportion of olivine, the mineral of which peridotites are made.

The core

The most widely accepted hypothesis on the composition of the core is that it is composed of iron, or possibly an alloy of iron and nickel. The basis of this suggestion is that if the core were made of iron, and the mantle of peridotite, the earth as a whole would be comparable in composition to meteorites, which are generally supposed to be the fragmented remains of another planet similar to our own.

There are different types of meteorite, but the main division is into stones and irons. Iron meteorites are more

often found than stony meteorites, but the latter are probably more common, and tend to be overlooked because they
look like ordinary rocks. The orbits of meteorites have been
studied and show that they originated in the solar system,
while their chemistry shows indications that they were formed
at a high temperature and pressure. It has been suggested
that they are the remains of a small planet which was once
situated between Mars and Jupiter. The stony meteorites
would correspond to the mantle of the disrupted planet and
the iron meteorites to its core.

There are no serious objections to the idea of the core of the
earth being made of iron, but there is no definite proof, and a
few scientists have tried to find alternative possibilities. One
such suggestion is that the core is similar in chemical
composition to the mantle but has been transformed into a
metallic material by the terrific pressure of the overlying
mantle. The pressure at the core-mantle boundary is about
1·3 to 1·4 million atmospheres. The fact is that no one really
knows what would happen to familiar materials if they were
subjected to such an immense pressure. It has been suggested
that the core might be composed of undifferentiated solar
matter, rich in hydrogen, although most seismologists are
rather sceptical about this idea.

Useful minerals

If we consider the smelting of ores, for example the copper
ores of Mansfeld in Germany, the melt generally separates
into three parts, a melt of heavy metals (mainly iron), a
matte rich in copper and iron sulphides, and a light scum of
the oxides of silicon, aluminium, calcium and magnesium
floating on top. In a way, this is like the structure of the

166

Figure 51. Kimberlite from the Premier
diamond mine near Pretoria, South Africa.
Kimberlite is a kind of peridotite.

earth itself, from its iron core to its crust of silicates, and just
as certain elements are concentrated in the melt and others in
the slag, so the various elements that occur in nature are
concentrated in different parts of the earth's structure. It was
proposed by Goldschmidt that the chemical elements
present on earth should be classified according to their
affinities, as shown in table 24. The elements of a particular
group often occur together in minerals; cobalt and nickel
occur in association with iron; copper, zinc and lead are
found together in sulphide ores; sodium, potassium, calcium
and magnesium are largely concentrated in silicate rocks.

It is hardly surprising that the different layers of rock
which we distinguish in the earth each contain a distinctive
selection of useful minerals. Thus, in the family of rocks with
the deepest origins, the peridotites, we find sources of
chromium and platinum. Diamonds are also formed at
great depth and pressure, as we have seen. Erosion of the
kimberlite pipes in which they occur has carried off many
diamonds to the sea, and along the coasts of south-west
Africa remunerative prospecting for diamonds is still possible. Special ships have been built to dredge up diamond-
bearing sands from the sea floor.

If we pass on to the crust itself we find that basalts are of
very little use in providing concentrations of useful minerals,
unless, as in the Cameroons, their weathering at the earth's
surface has left a residue of bauxite, the raw material from
which aluminium is obtained. The granites of the continental
crust, on the other hand, are richly endowed with mineral
concentrations. Tungsten and tin are found in veins and
dykes in areas of granite.

Other metals have a more complicated history, if one may
judge by their occurrence. Uranium for instance is a metal of

crustal origin but is sometimes found like tin in rocks of the granite family, sometimes like copper in sulphur-rich veins, or again sometimes found associated with vanadium or organic material in sedimentary rocks. What plant or animal could have concentrated vanadium and uranium in this way? Was it during its lifetime or after its death? It would be interesting to know, and research to find the answer to this question is at present under way.

Living organisms have intervened in the mineralogical history of the earth with more effect than one would suspect at first glance. Many concentrations of iron minerals are due to the action of bacteria in marshes or on the ocean floor, and those of the Krivoi Rog in Russia probably have this origin although they have subsequently been metamorphosed. Ultimately the very substance of living organisms decomposes, giving us coal and oil, and thus justifies the separate class given to carbon in table 24.

Certainly, for an element to be of practical use it must not

Table 24 Classification of the principal elements

after Goldschmidt and Mason

Atmophiles	reach a relative maximum in the atmosphere. They are chemically not very reactive. Nitrogen, argon, neon, and other rare gases.
Biophiles	reach a relative maximum among living organisms. Carbon.
Lithophiles	have a high affinity for oxygen, and a relative maximum in the earth's crust. Oxygen, silicon, aluminium, sodium, potassium, calcium, barium, magnesium.
Chalcophiles	have high affinity for sulphur, and probably attain a maximum in the mantle. Sulphur, selenium, arsenic, antimony, bismuth, copper, silver, zinc, mercury, lead.
Siderophiles	have a low affinity for oxygen and for sulphur, and a probable maximum in the earth's core. Iron, cobalt, nickel, phosphorus, gold, tin, germanium, platinum and similar metals.

only be present, but existing in such a form and concentration as to make exploitation worthwhile. Table 25 lists these desirable concentrations and compares them with the actual average concentration available. The comparison indicates the degree of enrichment necessary, and one may see that it varies between four and 500,000 times the average crustal concentration.

The exploitable reserves of the world's principal elements have been estimated by mining engineers and economists, and by dividing them by the annual consumption or production one arrives at an estimate of their assured period of

Table 25 Actual and exploitable contents of the useful elements

after Bartels, Jaeger, and Mason

| | Concentration in p.p.m. | | |
	necessary minimum for exploitation	actual mean content in the earth's crust	ratio of minimum to mean
Aluminium	300,000	81,300	4
Iron	300,000	50,000	6
Manganese	350,000	1,000	350
Chromium	300,000	200	1,500
Zinc	40,000	65	615
Nickel	15,000	80	190
Copper	10,000	45	220
Lead	40,000	15	2,700
Tin	10,000	3	3,300
Uranium	1,000	2	500
Tungsten	2,000	1·000	2,000
Antimony	100,000	0·200	500,000
Gold	6	0·005	1,200

p.p.m. = parts per million = grams per metric ton

existence. For many substances like copper, lead, and even oil, which are by no means the least important, the answer is astonishingly small – from twenty to one hundred years. We need not worry that supplies will be exhausted, however, for progress in the methods of prospecting and extraction will undoubtedly permit the utilisation of poorer and poorer sources, of which the reserves are more than adequate.

It is clear that a good knowledge of the distribution of elements over the earth is very valuable, and our interest in this is governed as much by practical considerations as by academic interest.

5 The earth's origin

The age of the earth

The more that man has learned about the earth, the greater has been his estimate of its age. The ancient Semitic peoples could remember twenty or thirty generations back from themselves, and through faith in the literal interpretation of the Bible, the Christian world believed for a long time that the earth was not more than five or six thousand years old. The earliest discoveries of the pioneer geologists soon showed, however, that the earth must have had a much longer history than a few thousand years. In the nineteenth century a figure of eighty million years was proposed, derived from the supposed time which a molten earth would take to cool to its present temperature. At the present day, radioactivity has given us new methods of dating, and so have astronomy and biology. Nineteen different methods taken together provide a very coherent picture (table 26).

The methods which make use of radioactivity are by far the most accurate, and they enable us not only to measure the age of the earth, but also to establish a time scale to measure its subsequent history. How are we able to state these ages? We make use of the fact that some of the isotopes of naturally occurring elements are radioactive, and transform continuously into other elements. The rate of transformation is a constant for each radioactive isotope and the rates at which the atoms change are known through laboratory experiments. For example, the isotope of uranium with an atomic weight of 238 changes to lead 206 at such a rate that 50 per cent of the original uranium would have changed to lead in a period of 4,400 million years. Knowing this, and comparing the number of atoms of radioactive element left with the number which have been produced by its

Table 26 Dates of origin of some cosmic and terrestrial phenomena

after Jeffreys and Cailleux

Event	Estimated date in million years Max.	Min.	Remarks source, method, etc.
Expansion of the Universe	14,000	5,000	Velocity of the nebulae
Formation of the sun	?	20	Solar radiation
Formation of the planets	4,200		Lead in meteorites
Formation of the earth	?	20	Temperature of the crust
Formation of the earth	4,500		Lead minerals
Very close presence of the moon	4,000	3,300	Earth's rotational deceleration
Formation of the earth's crust	5,000	?	Uranium^{235}
Formation of the earth's crust	4,000	?	Lead/uranium ratio
Crystallisation of the rocks		3,360	Radioactivity in rocks
Consolidation of the continents	4,200	2,600	Speed of accretion
Formation of the oceans	?	350	Sedimentation
Formation of the oceans	?	180	Sodium in the oceans
First living organisms	5,000	2,000	Longevity of the orders, etc.
First animals	5,000	1,800	Scales of psychism
First fossils preserved		2,600	Algae (Canada)
Speciation	?	2,200	Number of species
Single-celled animals	4,000	2,000	Scale of anatomical evolution (Huxley)
First fossilised animals	?	2,200	Number of species
Many celled plants	?	950	Number of cells

transformation, it is possible to obtain the age of the mineral under consideration. The principal radioactive transformations which are used for measuring ages are:

uranium 238	$\longrightarrow$	lead 206
uranium 235	$\longrightarrow$	lead 207
thorium 232	$\longrightarrow$	lead 208
rubidium 87	$\longrightarrow$	strontium 87
potassium 40	$\longrightarrow$	argon 40

We can thus measure the age of any mineral which contains one of the radioactive elements, and since potassium in particular is widely distributed there are many rocks for which it is possible to obtain an accurate age. The oldest rocks which have yet been found, in the Kola peninsula, are 3,500 million years old.

The figure of 3,500 million years sets a minimum age to the earth, but it is possible to go further back than this in time by studying the isotopes of lead. The relative proportions of these isotopes have been changing during the course of time by the addition of lead 206 and 207 from the transformation of uranium. By measuring the isotopic ratios of lead minerals whose ages are known it is possible to calculate the length of time during which this process has operated within the earth. The method, although not very accurate, yields a figure of about 4,500 million years, and provides our only direct estimate of the age of the earth. A similar method can be applied to meteorites with greater accuracy, and the ages of these are also found to be about 4,500 million years.

The most remarkable aspect of these results is that they are so close to the astronomers' estimates of the age of the universe obtained by a quite different method. Astronomers have found that the universe is probably expanding, and the rate of expansion can be estimated from measurements of the displacement of spectral lines in the light from distant stars. The velocity of recession of stars at a distance of 3×10^{21} kilometres, for example, is about 20,000 kilometres per second. Assuming that the expansion of the universe is uniform, all the matter in the universe would have been in the same place about 5,000 million years ago.

We must not, however, repeat the mistake that the ancients made. The 5,000 million years that we attribute to the universe is not necessarily the time since its origin, but merely that of the expansion which we have been able to apprehend. Who can tell whether perhaps this same universe did not have another history before the present expansion, which our successors, better equipped than we are, will be able to unravel?

Origin of the solar system

The earth and the other planets (except perhaps Pluto) have so many features and characters in common that one cannot doubt that they had the same origin. These common characters, which must be taken into account in any explanation of the origin of the planets, may be summarised thus:

1 All the planets rotate about the sun, and describe almost circular orbits, in contrast to other heavenly bodies which often describe very elongate ellipses. These orbits are all more or less in the same plane, which is inclined at approximately six degrees to the equatorial plane of the sun. All the planets orbit in a similar direction, rotating about their own axis in this same direction, except for Uranus whose equator is inclined almost at right angles to the others.

2 The distances of the planets from the sun form a series in which the separation between the planets increases with

their distance from the sun in a very nearly geometric progression. This law was discovered by Bode in 1772, and it predicted the presence of another planet in the empty space between Mars and Jupiter. Thirty years later this region was found to contain not a planet but a whole swarm of little planets, or *asteroids*.

3 This particular region separates the planets into two clearly distinct groups. Mercury, Venus, Earth and Mars are all small, fairly dense, and close to the Sun; they rotate slowly on their own axes, and possess few satellites. The other group, the light planets Jupiter, Saturn, Uranus and Neptune are all large and far away; they rotate quickly upon their own axes and have many satellites. Pluto, the most distant planet, seems to form an exception. Like the earth, the denser planets are probably poor in gaseous elements, whereas the other group are relatively rich, especially in methane, ammonia, and probably also hydrogen.

4 While the sun alone forms 99 per cent of the total mass of the solar system, it contributes only two per cent of the angular momentum, and the planets 98 per cent. Angular momentum is a function of the mass of a body, its velocity, and its distance from the axis about which it is rotating. Jupiter carries a large proportion of the angular momentum of the whole solar system.

Since the eighteenth century especially, the problem of the formation of the earth and planets has fascinated scientists and mathematicians, and there have been more than thirty hypotheses proposed to account for it. All of them are open to criticism and none are completely convincing, but several of them no doubt include at least some measure of truth. Table 27 summarises some of the most plausible. All these hypotheses have a common starting point for the substance of the planets as small particles. These particles must have been either solid or gaseous. They certainly could not have been liquid because they would soon have evaporated in the near vacuum of space. The first stage in the process would have been the agglomeration of the particles. All the authors are in agreement about this process and that in it, heat was evolved. Thus a newly born planet would warm up as long as the heat was not dissipated too quickly by radiation from its outer surface. It is here that the different hypotheses diverge: were the original particles gaseous or solid? Were they hot or cold? Did they come from the sun or from other cosmic matter scattered throughout space? It is impossible to decide upon the last question because in either case the chemical composition would be the same. One can arrange the possible answers to these questions in two broad groups: those described as uniformitarian, which envisage the formation of the planets by a continuous slow process, and those termed catastrophist, which imagine their formation to have been an unusual and violent event.

The first scientific theory, proposed by Kant and Laplace, was of the uniformitarian type. As Laplace saw it, a series of rings were progressively detached from a mass of hot gas, the future sun, each of which condensed to give a single planet, while the sun itself condensed from the gas remaining at the centre. This hypothesis has however several insurmountable difficulties, among which is the fact that it contradicts one of the great principles of nature, the conservation of angular momentum. The original mass would have had a certain momentum, and calculations show that a ring separating off from the outside would have a smaller angular momentum than the remaining mass. This is the opposite of

Table 27 Theories on the formation of the earth and other planets

	Particles	Nebulae	Planetesimals	Tides
Author, date	Kant 1755	Laplace 1796	Chamberlin and Moulton 1900	Jeans 1916
Catastrophic (Cat.) or Uniformitarian (Unif.)	Unif.	Unif.	Cat.	Cat.
Internal (Int.) or external force (Ext.)	Int.	Int.	Ext. tidal	Ext. tidal
Original material	Future sun	Future sun	Sun	Sun
Gas or solid	Particles	Gas	Gas then solids	Gas
Hot or cold	—	Hot	Cold	Hot
Originating event	Turbulence	—	Passage of star	Passage of star
Force	Mechanical	Mechanical	Mechanical	Mechanical
Original form	Unformed	Rings	Two protuberances	Filament, cigar
Orbit around sun	Badly explained	Postulated	Postulated	Explained

Double star	Physical chemical differentiation	Cosmic dust	Solar organisation	From a supernova	From twin planets	From turbulence and accretion
Russell 1921 Lyttleton 1936	Kuhn and Rittmann 1941	Weizsäcker 1944	Schmidt 1944	Hoyle 1944	Dauvillier 1947	Urey 1952
Cat.	Unif.	Unif.	Unif.	Cat.	Cat.	Unif.
Ext. tidal	Int.	Int.	Mixed	Ext.	Ext. tidal	Int.
Companion of sun	Sun or star	Cosmic cloud	Cosmic cloud	Companion of sun	Sun	Cosmic cloud
Gas	Gas	Dust + gas	Dust + gas	Gas	Gas	Meteorites, dust
Hot	Hot	Cold	Hot and cold	Hot then cold	Hot	Cold
Passage of star	—	—	—	Explosion	Passage of star	—
Mechanical	Mechanical + ionisation	—	Mechanical + radiation	—	Mechanical and electro-magnetic	—
Filament	—	Disc	Disc	Disc	Two opposing jets	Disc
—	Postulated	Postulated	Postulated	Explained	Explained	Pos

the actual case in which Jupiter, as we have seen, has much greater angular momentum than the sun.

It is very strange that it took almost a century before scientists seriously understood this difficulty. To escape from it, let us consider the earliest catastrophist hypotheses. Chamberlin and Moulton in 1900 imagined that a star had passed very close to the sun and caused two gaseous protuberances, one on either side of the sun, one of which gave rise to the lighter planets and the other to the dense ones. Condensation of these gaseous masses would initially have given rise to a large number of small planets, which they called *planetesimals*, followed by further aggregation into larger planets. It is easy to see that in the planetesimal stage, all the gases and vapours, including water, would have escaped because of the low gravitational attraction of the individual bodies. Consequently, the earth's oceans, its atmosphere, and the gases emitted from volcanoes all argue against this hypothesis.

Jeans in 1916 also rejected the intermediate planetesimal stage, and although agreeing with the idea of a disturbance of the sun by another heavenly body, saw it as causing a gigantic tide on the sun, which of course is gaseous, giving rise to a cigar-shaped filament which soon divided into planets. The difficulty here is to explain the very great distance between the light planets with their enormous angular momentum, and the sun.

In order to overcome these difficulties Russell and Lyttleton brought a third actor into the drama: another star which accompanied the sun. There is nothing exceptional in this, for ten per cent of the stars which we see are twins. When the heavenly body responsible for the disturbance peared on the scene, the intruder worked its effect upon the sun's double, causing a tidal disturbance which eventually gave rise to the planets: for this line of reasoning the planetary distances are on the whole correct. Looking into the sky today, however, where can we find that other star? There is only one sun. Lyttleton was able to show that under certain conditions of mass and speed, the intruding body could have deflected the sun's double, permitting it to escape from the attraction of its companion. At the risk of being irreverent we could say that the intruder had seduced the sun's wife, leaving the wronged husband to look after the children which she had had by the seducer. The only problem is that the condition of the children, the planets, is hard to understand. In particular, if they were born independently of the sun, why are they so regularly arranged around it? In addition, the physical properties of the gaseous matter making up the sun are such that any matter which became detached would tend not to condense into planets, but rather to dissipate itself in space. The same objection must, on the other hand, be applied to all the theories involving tides induced upon the sun.

During the last thirty years many new theories have been put forward and discussed, and interest in the older uniformitarian ideas has been renewed. Urey, for example, proposed that the planets were built up by the accretion of small particles or meteorites. The principal attraction of this theory is that it avoids the problems arising from the differences in chemical composition between the sun and planets. On the other hand, the earth is bombarded by about five hundred kilograms of meteorites every day, and although this seems a lot it would require three thousand million years to increase the mass of the earth by one ten thousand millionth, giving a layer only a millimetre or so thick. Even supposing that there had been a hundred or a thousand times more meteorites, they would still not have been enough to form a mass as great as the earth, unless the process took a much longer time than the known age of the universe. As for the gases, Urey would have them deposited as a hoar frost on the cold solid particles; but it is difficult to see why the frost should not have sublimed into the vacuum of space. Meteorites themselves contain very little gas.

The difficulty of retaining the gases, necessary to form the atmospheres of the planets and their oceans, is evident in all the other hypotheses which involve agglomeration at low temperatures, such as those of Weizsäcker and Schmidt. There are other difficulties too: the small angular momentum of the sun is not explained, and there are problems relating to the production of the heavy radioactive elements present in the earth. Such elements could have been produced only from the extremely condensed and hot material in the interior of stars. It seems inevitable that an earth containing such elements must have been derived from the sun or another star.

It was this which inspired Hoyle, the English mathematician, to revert to a theory similar to that of Laplace. He added to it a new idea, due to Alfven and Dauvillier: that of the magnetic field. The theory starts with a rapidly rotating cloud of gas, destined to become the sun and planets, contracting more and more; when its equatorial diameter becomes smaller than the orbit of Mercury some gas leaves the equator and forms an external disc, which rapidly becomes larger. The magnetic lines of force issuing from the central body (the future sun) in the plane of the disc twist spirally and tend to brake the rotation of the sun, while at the same time accelerating that of the external gaseous disc.

The angular momentum – the stumbling block of so many other theories – is naturally transferred by this means from the sun to the outer disc, that is, to the future planets. The gases of the disc then condense into the planets, the composition of which depends on their distance from the sun: nearby the temperature is high and it is principally the heavy atoms which condense, forming the heavy inner planets; farther from the sun, where it is colder, the lighter atoms condense in their turn to give the outer planets. One comes back here to the ideas of the Soviet astronomer Schmidt. This theory of Hoyle's is the most complete which has yet been put forward, and it is very probably a close approach to the truth.

The moon

Any theory which is advanced to account for the formation of the earth must also explain the presence of its satellite, the moon. That it has some relationship with the earth seems quite probable, especially since its density of 3·33 is the same as that of the upper part of the earth's mantle. At the present time the latter is 2,900 kilometres thick, and the moon's radius is only 1,730 kilometres. One attractive idea is that the moon originated from a tidal swelling on the earth. The astronomer Darwin, son of the illustrious biologist, and the great mathematician Henri Poincaré, have both calculated this to be a possibility. Since then, however, more detailed calculations have revealed too many difficulties: the amplitude of the swelling would be too small, or the moon would fall back to earth as soon as it had been cast off, or on the contrary it might escape altogether into space.

Meanwhile, we must discard one false idea, that is, the

notion that the Pacific Ocean, the biggest of all depressions on the earth's surface, is a scar left by the moon when it separated. In fact if such a separation did take place it could only have been when the earth was liquid, and once the swelling of the future moon had been ejected the remaining cavity in the molten mantle, if there ever was such a thing, would have closed of its own accord and the earth's surface evened out again. Also, the Pacific Ocean has a volume thirty thousand times less than that of the moon, and this is small for the hole which is supposed to have been left!

Consolidation of the earth

Was the earth formed as a hot fluid body or a cold solid? Its flattening shows that it either is or has been deformable, and there are numerous indications of its fluidity. The remarkable internal superposition of layers, of different chemical compositions as far as is known, arranged in order of density, would have been facilitated and speeded up if the various atoms and compounds could circulate in convection currents. A particularly remarkable fact is the upward concentration of radioactive elements in the crust. From present-day measurements of heat flow in the rocks it would require twenty kilometres of granite to explain the total heat supplied to the surface of the continents, such is the concentration of heat-producing radioactive elements in the granitic rocks of the crust; the remainder of the earth has a considerably lower concentration. In 1925 the mineralogist Goldschmidt showed how this distribution could have come about: atoms of uranium and thorium cannot enter the crystal structures of olivine and the other minerals which probably make up the rocks of the mantle, but they can very easily occupy a place in the structure of zircon, a mineral which occurs in granites. Such a process of concentration would have been impossible without by-passing an immense number of other atoms, a gradual movement which would only take place under conditions of mobility in the interior of the planet, in the past at least. The core, in fact, has remained mobile right up to the present day, as we can see from its inability to transmit secondary earthquake waves. Higher up, at the boundary between crust and mantle, there is matter which is pretty close to being liquid, because it frequently melts sufficiently for lavas to be poured out at the surface.

An important point to remember about the production of heat by radioactive elements is that radioactive atoms can only furnish their energy by disintegrating, and at a rate which is known from laboratory measurements. It is a simple step to apply the known rates of disintegration to the past, and it may be calculated in this way that the earth has already used half its supply of radioactive fuel. Four or five thousand millions years ago when the planet took shape it contained twice as many radioactive atoms as it does today. In those early days the emission of heat must also have been twice as great, which further favours an original molten state.

It is thus extremely probable that the earth has passed through a liquid or gaseous state, at least in its early days. All theories which postulate its birth from the sun or a star at very high temperatures would allow this. Even in other theories which see its origin in cold dispersed particles there is the possibility of its having gradually heated up through the energy of gravitation and by radioactivity. The earth would thus have at least arrived at a partially molten state and no doubt with a gaseous atmosphere as well.

A fundamental problem, that of escape velocity, is presented immediately by the mention of gases. This has been made more familiar to us through work with space rockets and artificial satellites. Gaseous particles, whether ions, atoms or molecules, are in a state of continuous agitation, all with different velocities often exceeding several kilometres per second, depending on the temperature of the gas. The higher the temperature, the greater the velocity of the particles. The speed of the particles also depends on their mass, the lightest particles moving most rapidly.

A gas which is enclosed in a fine grained sediment will remain there. If it is free in the atmosphere however, what then? For any heavenly body there is a characteristic quantity called the escape velocity which depends on the gravitational attraction exerted by the body. For the earth itself this is 11·2 kilometres per second, and any particle which exceeds this velocity in a direction away from the earth, and does not collide with any other particle, will escape from the earth's attraction and fly off into space. This applies to large objects such as interplanetary rockets, and must apply equally to small objects such as the atoms of the lightest elements, hydrogen and helium. This explains very clearly why these elements, predominant in the sun where they are held by the attraction of its great mass, are much more rare on the relatively small earth; the atoms of these elements must have escaped in great numbers from the earth and other dense planets early in their history. On the other hand, the escape of these elements and other light particles is never complete because of the population effect. Some individual particles always have a lower velocity than that required for escape, or do not travel in a direction away from the earth, collide with other particles and are deflected back towards the earth, or combine chemically with other elements to form large slower-moving molecules.

Let us then consider the development of the earth from its origin, allowing it an initial composition the same as that of the sun, and following Dauvillier's suggested sequence of changes. Right at the beginning the earth's mass was perhaps twenty times greater than it is now because of the high proportion of hydrogen and helium which it contained, and the escape velocity could have been around seven kilometres per second, appreciably less than at the present. The light particles would have tended to escape more readily, especially in view of the greater temperature, something close to that of the sun, and perhaps as high as 4,000°C. Elements with atomic weights up to twenty would have escaped in large numbers, including carbon, nitrogen, neon and the lighter groups of atoms. Some light elements would have been held by combination with heavier elements such as oxygen, which may be bound to heavier elements in oxides, silicates and many other types of compound.

Nevertheless the gaseous mass would have been cooling quickly, and kept in motion by intense convection currents. Towards 3,000°C certain substances would have started to liquefy, with iron first, being heavy, condensing towards the centre where it started to form the core, including the inner core, which may not have become differentiated at this stage. In their turn the silica and the metallic oxides liquefied, giving the molten forerunner of the mantle. As the temperature decreased so the radiation of heat into space decreased. and the cooling process slowed down.

Between 1,500°C and 800°C an essentially new scene w enacted: the solidification of the crust. We may here hesi a little between basalt and granite. Basalt is the fir solidify, but the solid form is heavier and would tend to sink in the hot melt, where it would melt again and the upper part be enriched with granitic constituents. It is thus possible that granite formed the first persistently solid crust on the surface of the silicate melt.

Meanwhile the atmosphere had been building up bit by bit, and was at that time composed of water vapour, ammonia, and the oxides of carbon. If all the water in the present oceans were concentrated in that atmosphere it would have exerted a pressure 260 times that of the present atmosphere. Thick and heavy, it would have tended to make climates more uniform. However the slightest heterogeneity would have been enough for the first solid crust to appear in winter rather than summer, and at the poles rather than the equator. From this beginning the crust would have grown and thickened, gradually extending over the whole surface of the earth, and persisting regardless of the season.

At the state we have just arrived at, the new solid crust was, however, an obstacle to the exchange of heat by convection currents, and subsequent passage of heat to the surface was by the process of conduction, which is a good deal slower. The cooling of the outer crust speeded up, and that of the inner regions of the earth slowed down. Rittmann has calculated that there could have been a possible ten kilometres thickness of solid crust by the time the temperature was down to 700°C.

Such is the history that one can infer, if the earth is assumed to have started life as a mixture of hot gases similar o that of the sun or another star. If, however, it originated the aggregation of cold particles, the temperature must e risen gradually, as we have already seen, and the earth, s outer layer, been able to liquefy. From this point it could have evolved in accordance with Dauvillier's hypothesis, allowing some small modifications.

Origin of sea water

Taking up our planet where we left it, with both the solid crust and the atmosphere continuing to cool, the next major change would occur at a temperature of about 374°C, the critical temperature of water, when the steam in the atmosphere would condense to a liquid; the critical pressure of steam is in effect only 226 atmospheres, lower than the 260 postulated to exist at that time. The difference, thirty-four atmospheres, corresponds to a layer of 350 metres of liquid water. This must have condensed as rain, starting in the colder regions around the poles. For the first time ever there was water streaming about the crust, corroding it, and gathering together in depressions forming the first seas. This resulted in the first centres of excess weight on the young crust, the first isostatic inequilibria. Towards the equator the temperature would have been slightly above 374°C to start with, and there the water would have boiled and returned to the atmosphere. The external temperature continued to go down. The boiling would have soon stopped, and with continuous condensation of water from the atmosphere, the heavy rain would enter the ocean and swell its volume. When the temperature reached 20°C or 30°C the earth's outward radiation would have ceased to predominate over the heat received from the sun, and the temperature must then have stabilised not far above its present value. As far as one can judge from the geological evidence of the l[...] six hundred million years of sedimentation, there has b[...] no great change in the earth's climate as a whole durin[...]

period, although some areas which were once cold are now warm, and *vice versa*, and some areas which once had a heavy rainfall are now dry, and again the reverse is true. The critical temperature of 374°C would have been reached fairly quickly, perhaps 15,000 to 20,000 years after the origin which Dauvillier postulates, and a temperature of 25°C to 30°C after 60,000 years or so.

Had the oceans mustered their total volume by then? Some geologists would say not. According to them a substantial part of the water in the oceans would have been gradually supplied later from the earth's interior by volcanic eruptions and hot springs. Looking at the balance of things, one finds, however, that in the four thousand million years which separate us from the earth's origin, the water ejected from volcanoes into the air could have provided only five per cent by volume of the actual present day oceans. In addition, even if a little water does continue to come from the interior in this way, it is possible that a little oceanic water leaks back into the muds and consolidated rocks on the ocean floor, and may even be returned to the oceans if the water emitted from volcanoes comes in part from earlier rocks.

The water content of granites, basalts, and other igneous rocks is very low, 0·8 per cent to 1 per cent by weight. If one supposes that all this water were extracted from the basaltic oceanic crust, it would give a layer 180 metres deep, and that from the combined continental crust would give a depth of seven hundred to eight hundred metres; the mean of 450 metres is very low compared with the 2,630 metres of the present oceans (table 5). The crustal rocks are not a reservoir any great capacity. In the earliest times of our planet, with temperature at 1,500°C to 2,000°C the same material melted could have held a little more water, three or five per cent judging from laboratory experiments, but once the rocks solidified this water would have been lost. On the whole it seems likely that it did not take long for the oceans to attain very nearly their present volume.

The salt problem

It has long been known that fresh water, including river water, contains various salts in solution. These are responsible for the furring of kettles and the formation of scum during washing. When these dissolved salts were first analysed chemically it was found that they differed considerably from the salts in the sea. Since the rivers are the main source of material entering the sea it is very difficult to see why the compositions should be so different. The problem becomes even more difficult when we try to balance individual elements. As an example, let us consider sodium, which is the metal combined in common salt and in washing soda. The oceans contain 12,600 million million metric tons of sodium, and year after year the rivers carry in a further 156 millions. If the oceanic water did not contain any sodium to start with, and if none has escaped in the meanwhile, it would have taken only eighty million years at the present rate to build up the present concentration of sodium in the oceans. We know, however, that there are outlets such as coastal lagoons where the sea water evaporates and the salts are deposited from solution, later to be buried under other sediments as beds of rock salt. Some part of the salt which rivers transpo to the sea go into such deposits. Taking this into accou the estimated time for the build up of marine salts rises f 80 million to 180 million years. Even so, there is a good of the 4,500 million years of the earth's history still l

Figure 52. Mammoth hot spring in Yellowstone National Park, Wyoming, encrusted by salt which it has deposited. Most fresh waters contain some salts in solution.

Figure 53. Vapours rising from Virunga volcano
in Central Africa. Such vapours contain small amounts
of many substances, such as sulphur and chlorine,
and much of the chlorine in the sea may have
originated from submarine and terrestrial volcanic sources.

Table 28 Present day composition of sea water compared with substances brought down by rivers since the beginning (after Rankama and Sahama)

Element	Composition of eroded material g/ton	Present day concentration in sea water g/ton	Proportion in solution per cent
Chlorine	118·4	18,980	10,074
Bromine	0·97	65	6,687
Sulphur	312	884	283
Boron	1·8	4·6	256
Sodium	16,980	10,561	62
Iodine	0·18	0·05	28
Magnesium	12,540	1,272	10
Potassium	15,540	380	2·4
Calcium	21,780	400	1·8
Iron	30,000	0·02	0·00007
Aluminium	48,780	1·9	0·004
Silicon	166,320	4·0	0·002

In order to clarify the picture it is advisable to ignore sedimentary processes and concentrate on the point of departure, the igneous rocks, and the point of arrival, the sea. Table 28 lists the actual content of the most important dissolved elements in sea water, and in the source rocks from which they have been carried by the rivers over a long period. The result is very interesting. For the four elements, chlorine, bromine, sulphur and boron, it is found that sea water contains a larger amount than the rivers could possibly have delivered. There must thus be another source, and volcanism is the obvious answer. These elements, among others, are particularly common in volcanic emissions and fumaroles. For all other elements the difference lies in the opposite direction, there is less in the sea than the rivers would have brought There must be loopholes somewhere, and we have two immediate possibilities: living organisms and the deposits of sediment on the sea floor.

Calcium, for example, forms part of the shells and skeletons and bones of a host of marine organisms. Silicon must also be considered in this connection, but the animals which use it, such as certain sponges, are considerably less abundant, and the remainder of the silicon must be lost in the form of sediment. This remainder is enormous, to judge from table 28. Aluminium must certainly be incorporated directly into rocks on the ocean floor, because it is not found in living organisms except in very low quantities, and is not very abundant in sea water either. This applies to a greater or less extent for most other metals.

Living beings constitute only an infinitesimal portion of the sea, and the proportion of elements actually incorporated in them is minute. On the other hand the amount of material incorporated in the shells of organisms which have died and been incorporated into the sediments on the sea bottom is quite considerable. A high proportion of calcium and silicon 'disappears' in this way.

Origin of life and atmosphere

Returning now to the early period, when the temperature was about 200°C and the atmosphere was composed chiefly of water vapour, carbon dioxide and ammonia, the surface waters of the oceans must have contained both carbon dioxide and ammonia. These substances contain all the ma⸍ elements needed to form the proteins and other material⸍ which living things are made, that is, carbon, hydrogen, gen and nitrogen. One can only speculate as yet on th⸍ in which the simple molecules were joined together ⸍

the more complex substances of living organisms, but some simple experiments have already provided important clues. It has been found possible to synthesise quite complicated organic molecules by the action of electrical discharges on mixtures of carbon dioxide, water, methane and ammonia. In nature, of course, the place of the electrical discharges would be taken by lightning. At some stage, molecules capable of reproduction must have been formed, and the first very simple living organisms were born. The temperature may then have lain somewhere between 30° and 100°C, probably at the lower end of the range. By analogy with natural processes and laboratory experiments, it is possible to follow most of the likely steps in the origin of living organisms.

Like certain living bacteria, some of the early organisms were probably able to break down ammonia from the sea water and release nitrogen and oxygen into the primitive atmosphere. Others would oxidise the ammonia, while some, capable of assimilating carbon dioxide, retained the carbon and set free oxygen. Thus the early living organisms may have played an important part in the development of our atmosphere of nitrogen and oxygen. A fortunate consequence of the presence of free oxygen is that in the highest layers of the atmosphere it forms ozone, and this gas absorbs ultraviolet radiation from the sun which would otherwise destroy life. In the three thousand million years or so which followed, living organisms have spread out and diversified, occupying not only the sea but the land and air as well.

6 Movement of the continents

Have the continents always been where they are now, or have they moved across the surface of the earth? At first sight nothing seems more firm and immovable than a rock across which the sea is breaking, but many rocks reveal a remarkably complex history when their fossils are examined and their structures are unravelled. Palm trees once grew in Paris and in London, and magnolias in Greenland, and glaciers once stretched over Brazil and the Congo. How could this have been possible? Have the poles and equator changed places, or are the continents themselves in movement? Let us see what various authorities have to say on the subject.

The astronomers' opinion

At the present time the poles are known to wander, and thanks to very accurate measurements of latitude by numerous observatories we are well acquainted with this phenomenon. Each pole describes irregular concentric curves or circles which make one think of an unskilled childish drawing. It takes approximately 433 days to complete a cycle and a square of side exactly thirty metres (a quarter of the area of a football pitch) is enough to enclose the total area of movement. Astronomical theory takes these movements into account, but they are clearly too small to have had an effect on the climates in various regions of the earth.

The question nevertheless remains of whether there would an observable displacement in a particular direction after rge number of cycles. The answer is not yet known for n because the century during which we have been mak- servations is too short a time. The scale of geological s incomparably greater, and the question remains the time being.

The mathematician's opinion

The mass of the earth is enormous, and its rotation upon its axis at great speed produces a large angular momentum. Any force tending to change the position of its axis of rotation would thus be strongly opposed. This principle is employed in the gyroscope, and explains its use in ship and aeroplane compasses where it takes up a fixed direction in space.

On the surface of the earth, however, the existing pattern of the continents ensures an unequal distribution of mass. In theory this could cause a polar displacement of slightly more than 10,000 kilometres over a period of perhaps 200 million years. The path of the displaced pole has been calculated by mathematicians; to start with it follows the 140° W meridian, and then runs fairly closely along 120° W, which we have seen to be in one of the earth's planes of symmetry. This is hardly surprising since both have been determined from the same geographical data. Although we do not know whether such displacements have taken place, we may at least conclude that if there have been important polar wanderings in the past they will have been very slow.

Evidence of magnetism

Nature has distributed little magnetic compasses throughout the rocks in the shape of the magnetic minerals, and the study of these has led to many interesting results. Magnetic minerals may be found particularly in volcanic lavas and fin grained sedimentary rocks (once muds), and the orientati and intensity of the magnetism in these rocks can be mea ed. The intensity in lavas is generally high, but in sedir it is a good deal weaker and therefore less reliable. C

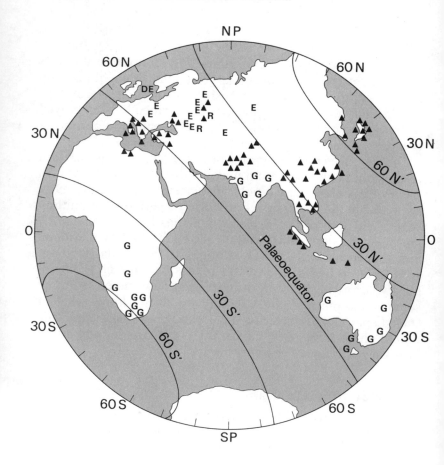

Figure 54. A reconstruction of the climate in Permian times (250 million years ago), based on the evidence of salt deposits (E), coral and other reefs (R), desert sandstones (D), fossils of warm-water organisms (triangles), and glacial features (G). An attempt has been made to draw 'palaeolatitudes' but it is not really possible to reconcile the apparent distributions of hot and cold conditions.

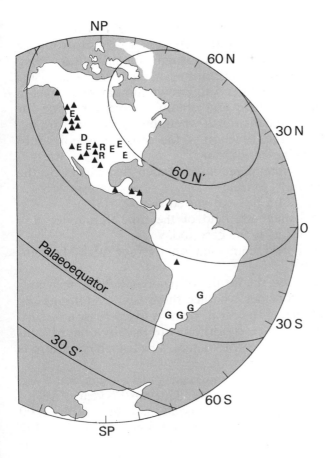

the essential precautions in making such measurements is to eliminate the effect of lightning strikes, which superimpose new magnetisation and mask the original magnetisation being investigated. By making several measurements in successive flows one can allow for the well known annual variations in the earth's magnetic field and obtain an average value for the direction of magnetic north when the rocks were formed. This will indicate the direction of the geographic pole, subject to the following conditions:

1 The minerals were magnetised under the effects of the magnetic field of the time
2 This magnetisation has not varied since
3 The principal magnetic field of the time dominated any secondary ones, as is the case today
4 The principal field coincided on average with the geographic north.

Before we can use the results of palaeomagnetic measurements we must consider whether these four conditions are adequately fulfilled. In general, there is every reason to suppose that the first three conditions are usually obeyed, and anomalous results are easy to detect. The fourth condition is more difficult to test.

Has the position of the magnetic pole always coincided with that of the geographic pole? During the years 1929 to 1945 the two increased their separation from 1250 to 1300 kilometres, a variation three thousand times greater than the greatest deviation of the geographic pole from its mean position. The magnetic pole of the southern hemisphere moved kilometres in net displacement between 1909 and 1960, which is a rate of fifteen kilometres a year. At this speed it move round the world in under 3,000 years, a mere in terms of the history of the earth. It is intriguing that its general displacement, with some deviations, has not followed any local fold directions, nor a meridian, nor a line of latitude. However, when we consider the variation in the position of the magnetic north pole over the last 7,000 years, as indicated by the directions of magnetisation in archaeologically-dated objects, and the historic flows of volcanoes such as Etna, we find that it coincides with the position of the geographic pole, and is rarely farther away from the geographic pole than its present position. This is a very encouraging result, for it would mean that we can rely on palaeomagnetic measurements to tell us the position of the geographical poles in the past, assuming, of course, that sufficient samples are measured to average out the minor fluctuations of the magnetic pole about the geographic pole. Let us see what results can be obtained.

The most remarkable result was quite unpredicted. In a suite of lava flows or a succession of sedimentary layers there are periodic reversals of the direction of the magnetisation, although its intensity remains the same. About every one, two or three million years everything that was north becomes south, and vice versa. With some lavas this has been proved to be due to a subsequent change in the physical and chemical properties of the magnetic minerals, but in other cases the sole explanation is that of an actual reversal in the earth's magnetic field so that the internal currents which govern and produce it must also have been reversed.

One practical consequence of this is that in reconstructing the positions of the magnetic poles one treats all the result together, without worrying about north and south. By c vention it is the position of the pole in the northern h sphere which one records.

If we first take as an example measurements made

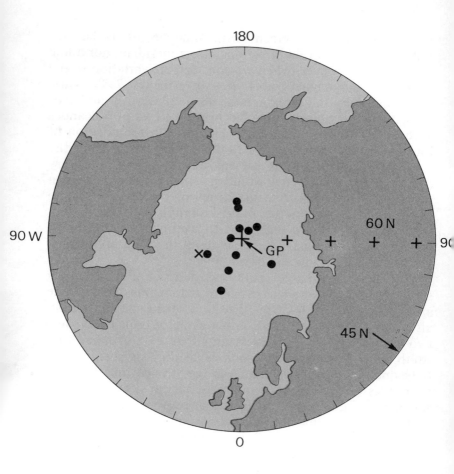

Figure 55. In the Northern Hemisphere the position of the magnetic pole as indicated by the residual magnetism of various materials up to 7,000 years old. X marks the position of the present geomagnetic pole; the geographic pole is indicated by the letters GP. The magnetic pole is never very far from the geographic pole.

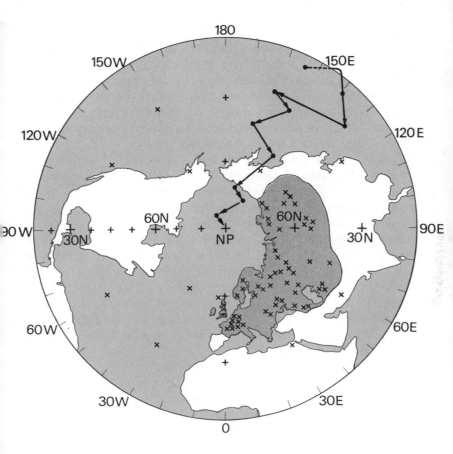

Figure 56. Apparent magnetic polar movement relative to Europe during the past 500 million years. The pole appears to have moved from a position in the Pacific area to its present position via the eastern tip of Siberia.

Antarctic continent, it is found that 600 million years ago the pole was located in the Pacific towards latitude 3° N and longitude 107° W, not far from Panama; 70 million years ago it was very close to the pole of the present principal field; 35 million years ago it had moved slightly away again, and a few tens of thousand years ago it coincided exactly with the present position. Results of this kind have been obtained from other areas besides Antarctica, and in every case the position of the pole appears to have moved. The obvious inference is that the poles themselves have moved, a pheno-menon described as *polar wandering*.

If we now gather the results from all parts of the earth we find that for the last sixty million years or so all the pole positions obtained are fairly close to the present geographical pole. Before this time, however, we find that there are differences. If we plot the various positions of the magnetic pole on a map, and join them up to give its path through time, it may be seen from figures 56 and 57 that while measurements from Europe give one path, those from North America give another, and there is yet another quite different one from India.

This result does not tell us definitely whether or not the poles have moved, but it does tell us that polar wandering alone will not explain the directions of magnetisation of rocks, for it shows that the continents themselves have changed their relative positions. This discovery, made gradually during the last ten years as more and more palaeomagnetic results were obtained, brought about a great awakening of interest in the theory of *continental drift*, an which was popular 40 or 50 years ago but came under s criticism around 1930 and was abandoned by many sts.

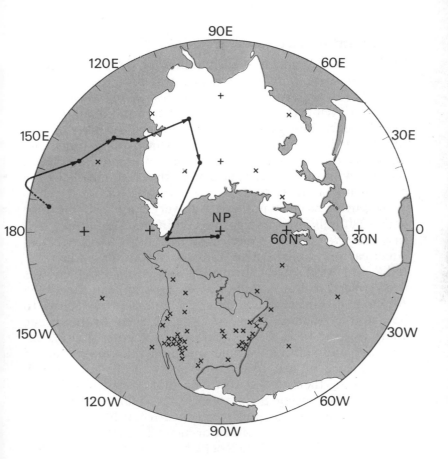

Figure 57. Apparent polar movement relative to North America during the past 500 million years. The pole appears to have moved from the Pacific towards its present position via Korea and northern China.

Continental drift

In 1658, R. P. François Placet published in Paris a work called: *Wherein it is demonstrated that prior to the Flood, America was not separated from the rest of the world.* Later authors developed similar ideas, but it was not until 1912 that Wegener drew serious attention to the possibility of continental drift. Wegener, who was destined for a tragic end on the ice in Greenland where he died of hunger, developed the idea with such vigour and ability that his name has been permanently attached to it.

Wegener's basic hypothesis was very simple. Some time about 280 million years ago the continents were united in a single block: South America slotted into Africa; North America was joined on to Europe; Antarctica, Australia, India and Madagascar formed a single block which was itself welded to South America. A crack appeared between Africa and America, and later on the fissure opened to become the Atlantic ocean, and the other continents broke apart in their turn. Being made of light crust floating on a denser and sufficiently fluid medium. they moved under the influence of two forces: the effect of the tides which was drawing them to the west, and that of centrifugal force theoretically causing them to move gradually towards the equator. Thus the Americas separated from the Old World. The resistance of the Pacific floor against their leading edges provoked the folding of the Rocky Mountains and the Andes. On the other side of the oceans, in the rear of the continents, island arcs represent the remains of the material of the continents which they could not retain in their drift to the west.

Wegener certainly cited many facts in favour of his

hypothesis, among which there is the astonishing fit of the coastlines of South America and Africa, and a multitude of resemblances between rocks, fossils, fold directions and faults in regions which are separated at the present day but would once have been united.

The hypothesis coordinated many otherwise scattered facts into a logical pattern. It provided for movement where previously all had been considered immobile, and in the years between 1920 and 1930 it enjoyed an immense success. Argand, a Swiss geologist, trained originally as an architect, gave a description at the 1924 International Geological Congress of the folding of the Alps as they were pressed in a vice between the two blocks of Africa and Europe. At about the same time Molengraaff envisaged the mid-Atlantic ridge as the former frontier between the Americas and the Old World. Staub considered the original continents to have formed two large blocks, rather than one: Laurasia, comprising North America, Europe and Asia (except India); and Gondwanaland, comprising South America, Africa, Australia, India and Antarctica, with the Mesogeic zone between them.

All these variations, and there are many more besides, are attractive, but what is the fundamental value of the theory itself? Many of the details of Wegener's theory present serious difficulties, and especially his proposed mechanism. Centrifugal force would not be strong enough to move the continents, and in any case the latter are not nearly rigid enough to move as coherent blocks. Nevertheless, althou the theory cannot be accepted in its original form it rema an attractive idea, and many geologists have tried to fir alternative mechanism.

Figure 58. The Chevallier-Cailleux hypothesis on the origin of the continents. Above the molten silicates there formed a light, acid slag or crust (granite etc), which soon covered the entire globe. The globe, turning more and more slowly, changed shape and became less flattened, which cracked the crust. The positions of the cracks were governed by this mechanism, and correspond well with the middles of the present oceans. The next stage was when the continents gradually contracted while becoming folded, and at the same time the fissures – the locations of the newly-forming oceans – grew in proportion to their present size. There was a slight shifting of the continents, accompanied by some rotation. The diagram shows an intermediate stage in the process.

Old fissures that have already become widened, eventual position. The underlying material that has been brought to light is shown in grey.

+++ Ancient fissure in the Antarctic region.

New fissures, eventual position.

____ Position of fissures developing in the granitic crust. They are destined to change their position on the globe as the crust contracts and becomes displaced.

Centre of rotation of the granitic crust.

⟶ Relative movement that the material of the crust would have in relation to the underlying material if the latter did not oppose the rotation.

Present mid-oceanic ridge, depth < 2000 m.

=== Present mid-oceanic ridge, depth > 2000 m.

· Epicentres of oceanic earthquakes occurring at a depth of less than 60 km.

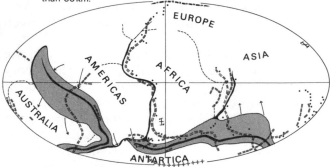

Figure 59. Convection currents are believed by many geophysicists to circulate in the earth's mantle, and have been widely invoked to explain structural features of the crust, such as the origin of fold belts (evidence of compression) and rift valleys (evidence of tension). Convection currents, if they exist, must move extremely slowly, perhaps only a few centimetres a year.

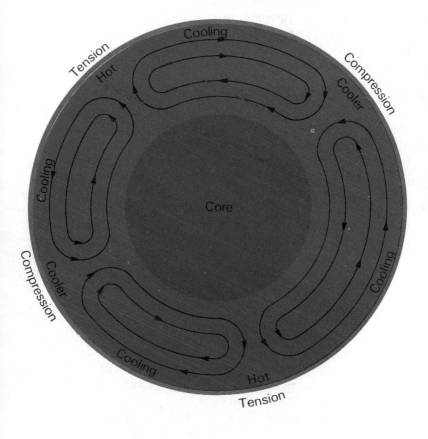

Convection currents in the mantle

We know that the mantle is rigid to abrupt stresses, such as earthquake waves, but the evidence of isostasy shows that it can react in a more supple fashion to more persistent forces. We are therefore justified in considering the possibility of currents within the mantle. In 1928 an origin was proposed for such currents, namely, temperature differences in the mantle, which could give rise to a stirring effect, or convection. A current may be envisaged as flowing upwards from the base of the mantle towards its boundary with the crust, turning outwards and flowing parallel to the surface, and then turning down into the depths again when it encounters an opposing current flowing towards it from another direction. Any such currents would of course be very slow, perhaps moving at a rate of only a few centimetres a year.

The theory was first applied by the Dutch geophysicist Vening Meinesz to explain the negative gravity anomalies, that is, the apparent deficits of material, which are observed beneath the deep oceanic trenches, and which Vening Meinesz was the first to reveal using gravity-measuring instruments mounted in a submarine. The idea was that beneath the crust of the deepest oceanic trenches in the upper layers of the mantle, two opposing currents meet and curve downwards, dragging the light crust with them. Alas, even the best scientists are not always fortunate with their theories, and studies of earth tremors and seismic explosions carried out on the best known submarine trench, that of Puerto Rico, have shown that there is no bulge of low-density crust into the mantle beneath it. The gravity anomaly due to another cause, the great thickness of ooze and other sediments which have accumulated in the trench.

other difficulty, although a less serious one, is that there are several secondary discontinuities in the mantle, between 400 and 1,300 kilometres down, which persist even though the convection currents are envisaged as flowing vertically through the whole thickness of the mantle.

Evidence from the ocean floors

The search for evidence on whether or not the continents have drifted has turned increasingly to the ocean floors, in the hope of finding out more about the history of the ocean basins. If the continents have not drifted, and the ocean basins are permanent features of the earth's surface, we might expect the sediments on the ocean floors and the fossils they contain, to reveal a history going back unbroken for hundreds of millions of years. The problem up to now has been the difficulty of obtaining samples from the great ocean depths, but this difficulty is being overcome.

The ocean floors can provide us with more direct evidence of their history, however, without the necessity of drilling into them. We have already seen that the earth's magnetic field has frequently undergone reversals, the north and the south poles changing places for no apparent reason. Recent surveys have shown that it is possible to measure the directions of magnetisation of the rocks underlying large areas and plot them on a map. Although the results obtained so far are not complete, some interesting results have already emerged. Surveys in different oceanic areas show remarkabl· similar patterns of normal and reversed magnetisation.

As an example, recent traverses across the mid-Atla ridge south-west of Iceland show alternating band· normally and reversely magnetised rock parallel to th of the ridge, the sequence of reversals showing rem·

bilateral symmetry on either side of the ridge. Since the rock at each point records the direction of the magnetic field at the time it was formed, it follows that the identical patterns on either side of the ridge represent two equivalent time sequences. There is no way in which this result can be reconciled with the hypothesis of permanence of ocean basins, and we can only conclude that the sequence of magnetic reversals is a record of the movement of the crust and mantle underlying the oceans, such as the convection current hypothesis would require. The magnetic measurements made so far show that the ocean floors are flowing away from the oceanic ridges at roughly equal rates on either side. The speed of movement can be estimated by correlating the magnetic sequence on the ocean floor with the directions of magnetisation in deep sea cores. The latter can be dated approximately from the fossils they contain, and the rate of movement of the ocean floors is found to be in the order of 1 to 5 centimetres per year.

The identification of the mid-ocean ridges with the upward currents of convection cells is supported by studies of the flow of heat through the rocks of the crust. The temperature in the earth increases with depth by 10°C to 50°C per kilometre, and consequently there is a steady flow of heat upward to the surface. The heat flow in any area can be calculated by measuring the thermal gradients in wells and boreholes, a correction being necessary to allow for the different thermal conductivities of different rock types. Not many measurements have yet been made on the sea floor, but ready it is apparent that the heat flow is greater under the ocean ridges than elsewhere. This is precisely what one expect if the upward movement of mantle material by ion was taking place under the ridges.

7 Mountain building

Volcanic and non-volcanic mountains

Having examined our planet as a whole, we are now in a position to take a closer look at the surface of the earth, and especially the question of how the great mountain ranges have come into existence. We have seen how there is a major division of the crust into the continents and ocean basins. There is the light continental crust, of granitic composition in its upper parts at least, thirty to thirty-five kilometres thick under the plains, and forty to seventy under the mountains. And there is the sub-oceanic crust, thinner and somewhat more dense, composed of five to seven kilometres of basalt.

There are many mountains scattered across the ocean floor, some surmounted by coral reefs, some exposed above sea level as islands, some beneath the surface as submarine hills. In addition to these isolated mountains, a great area of the oceans is occupied by the mid-oceanic ridges, forming as we have seen a continuous belt many thousands of kilometres in length. A common feature of both mid-oceanic ridges and isolated islands is that they are made of volcanic material.

Those parts of the continents with which we are acquainted are quite different: of the continental surface only one to three per cent bears volcanoes. Recently folded mountain belts and older folded rocks predominate, although the latter have often been reduced by erosion from their original mountain-ous relief to low-lying shields or platforms, particularly the very old ones such as those of Finland and eastern and central Canada. More rarely, a fold belt which has been worn down may be uplifted and redissected, as in Norway. Of course, it is quite possible for a region to contain both folded mountains and volcanoes, and this is the case in Japan, Asia and the Andes.

Every year the terrestrial volcanoes eject on average 1 to 1·5 cubic kilometres of matter, expressed as the equivalent volume of solid rock (see table 9), while 8 to 13 cubic kilometres of rock are eroded from the continents and carried out to sea. The difference, 6 to 12 cubic kilometres, would have been enough to fill the oceans in two hundred million years if no other force were in action, a mere fraction of the time that the earth has been in existence. What must have happened to prevent the seas being choked with sediment was the reincorporation of a good deal of it into the continents through folding.

A glance at the map is sufficient to show that terrestrial volcanoes often follow the line of recent zones of folding or fracture; the deformation of rocks obviously encourages the rise of material from the depths. Areas of frequent earthquake activity are often also rich in volcanoes but the relationship between earthquakes and volcanism is not as close as one might expect. The great expanses of ocean floor studded with volcanoes are essentially stable. It is true that there are earthquakes associated with volcanic eruptions, but such tremors are quite mild. Furthermore, huge earthquakes like the ones in Lisbon in 1755 and San Francisco in 1906 were not accompanied by any eruption. On the whole we may say that volcanic activity and earthquakes both indicate some instability in the earth's crust, but not in the same way; sometimes the two coincide, sometimes they do not.

Causes of mountain building

If we turn our thoughts to the tremendous landscape Alps or the Rockies, where the snow glistens all t round on the highest summits 3,000 or 4,000 met

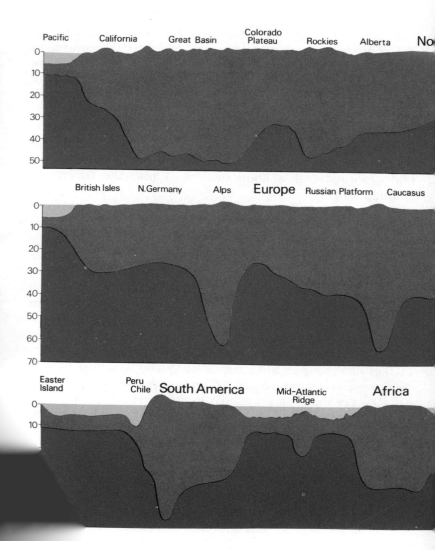

Figure 60. Profiles of the earth's crust (grey) and mantle (red) across the continents and oceans (vertical scale greatly exaggerated). Under every high area or mountain range the lighter crust projects downwards into the heavier mantle, thus preserving isostatic equilibrium.

merica   Wisconsin   Ontario   Appalachians   Atlantic

ergana Depression Pamirs Tarim Basin   Asia   Korea   Japan

Seychelles adagascar   Indian Ocean   Australia   Tasman Basin   New Zealand

sea level, it is difficult to realise that corresponding to each of these mighty ranges there is a root which penetrates more than fifty or sixty kilometres into the earth's interior. The higher the mountains stand above the plains, the deeper the root which projects beneath the crust. In projecting downwards, the comparatively light root surrounded by denser mantle material supports in some way the excess weight of the mountain chain elevated into the atmosphere. There is an equilibrium, an isostasy.

If we wish, however, to explain the actual existence of the mountain range we must also explain its root. We must also take into account the other characteristics of mountain ranges, and bear them constantly in mind: their formation on the borders between the continents and basins, of which the latter are generally marine; the fact that the most ancient massifs lie enclosed at the centre of encircling fold belts of younger rocks; the elongated form of mountain belts often with sheaves of more or less arcuate folds; the great thickness of the folded strata, often more than 10,000 metres. We have barely considered the last of these characteristics. It is quite remarkable that sequences of rocks one thousand or two thousand metres thick have never been folded, except as a part of the basement upon which they were deposited. Another fundamental feature is the sideways shrinkage which accompanies mountain building, due to the folding which takes place. This phenomenon is not always fully appreciated; it is often said that metamorphism has profoundly altered the nature of rocks in the older fold belts. In the field various ck types can nevertheless almost always be recognised in e of this, and one can nearly always deduce that the a have been remoulded, intensely folded, and severely ed in terms of the area of the earth which they occupy.

If a major part of the continents has shrunk, other parts of the earth must be under tension by way of compensation – places such as the ocean floor perhaps, and certain places on the continents such as the rift valleys of East Africa. Then again a mountain fold belt is usually between two hundred and one thousand kilometres wide, and one intuitively feels that the force which gave rise to it must have acted either along a front at least as wide as the belt itself or at some appreciable depth (one hundred to five hundred kilometres). The folding, however, affects only a small thickness of rock, the thirty to seventy kilometres of crust, and this encourages one to suspect lateral forces acting at a relatively moderate depth.

In addition to folding, mountain ranges show fissures, fractures, faulting and plutonic igneous intrusions, but the existence of these is easily explained by any of the theories in that they are all manifestations of forces causing deformation of the crust.

The characteristics described above are incorrectly stated by some authors to be characteristic of all folded mountain ranges. For example, it is true that in the majority of cases where the metamorphism has not been too great and one can assess the origin of the folded rocks they are marine sediments. This is not an invariable rule, however, and one finds that in the southern Andes and the interior of China the folded rocks are continental non-marine sediments. It is also said that the folding is cyclic, but this again is inaccurate as it is actually intermittent, and any hypothesis must be capable of explaining this. It would seem to follow quite simply as a consequence of the way in which stress builds up in the crust until the threshold of resistance is reached, when it is released in the fold or the fracture.

Figure 61. The Grand Canyon, Arizona. This immense gorge has been carved out by erosion over several million years. All the material eroded has been removed by the Colorado River and deposited farther downstream or in the sea. At this rate the oceans could be filled with sediment within a relatively short period of geological time if no compensating phenomenon occurs.

The contraction theory

Let us return to the scientific environment at the beginning of the nineteenth century. The constitution of minerals and rocks had just been established. Thanks to the increasing depths reached by mines it became known that although seeming cold at the surface the earth is hot inside; Laplace's theory invited us to view the earth as a body gradually cooling down. Thus it was natural for Elie de Beaumont to conceive in 1829 the theory of *contraction*. In modern terms it can be summarised thus: the crust on the outside was the first to cool, and did so quickly; beneath this, between seventy and seven hundred kilometres from the surface, the mantle had not yet finished cooling. It thus continued to reduce in volume, to contract, so that in three thousand million years its radius had decreased by about forty-two kilometres. The overlying crust became somewhat compressed, and it crumpled into folds. The shrinkage that it would have undergone during this period may be easily calculated:

$$2\pi R = 2 \times 3\cdot14 \times 42 = 264 \text{ kilometres.}$$

The circumference of the earth is thus reduced from 40,264 to 40,000 kilometres, a shrinkage factor of $1\cdot0015$. In relation to the degree of compression undergone by folded mountain chains, as evidenced by their severe folding, a factor of $1\cdot0015$ seems almost negligible. The recent Alpine ranges themselves could probably account for the whole 264 kilometres, and one is then left with no explanation for the shrinkage of all the older mountain belts. The theory of contraction must therefore be abandoned. It has also several other inconvenient features. The importance of horizontal displacements along faults is poorly explained, if it is to be taken as true that the decisive movement has been in towards the earth's centre. It does not really explain why the folds should have grouped themselves in ranges bordering a more ancient belt of folding, and one would rather have anticipated scattered wrinkles and puckers. The existence of zones under tension, such as the rift valleys, is inexplicable if the dominant movement has been contraction. As a final blow, it is not at all certain that the earth is in fact cooling. A large part of the flow of heat, under the continents at least, is due to radioactivity, and the minerals which provide this are definitely more concentrated within the crust than beneath it. It could even be that they are contributing heat to the layers beneath. We have more than enough here to renounce the theory of contraction by cooling, and to look for something else.

There is a modification of the contraction theory which assumes that the crust cooled more quickly than the mantle beneath, and that it was this which contracted, and not the mantle. Cracks appeared, forming more or less triangular fragments, or continents. These pieces of crust, however, had a greater curvature than the mantle, and in remaining in contact with the mantle they tended to fold along their edges. Unfortunately this modification still does not remove most of the difficulties associated with the contraction theory. It does not explain the size of the crustal shrinkage any more than the previous theory did, neither does it explain the existence of ancient fold belts in the heart of the continents. The effect could have played some part perhaps, but it was not by any means the most important.

Theories of vertical movement

Some of the theories which have been put forward require a vertical movement of the crust to initiate the process of mountain building. According to the *oscillation theory* of Haarmann, and several other theories, a series of parallel crests and depressions are formed in the land as a result of vertical movements of the crust. The strata which have been elevated tend to slip down into the depressions under their own weight, and they thus become folded.

The idea of folding through a downward slipping movement under gravity is common to a large number of theories. It has been envisaged as relevant to the Alps since 1892. In this range of mountains, and many others, sheets of rock, or *nappes*, can be seen to have moved almost horizontally over other rocks beneath them. The problem in these instances is to decide whether they have been pushed by some external force up a slightly inclined plane, or whether they have slipped down it under their own weight. In some cases in the Swiss Pre-Alps, for example, it has been proved that they slid down. Clear examples are rare, however, and an upward slide under some pushing force can often be just as easily envisaged, if not more so. In nine out of ten cases the base of the nappes or rock sheets is at present inclined in the opposite direction to the slope which they would have descended, a slope in the order of 10°. It would be possible to explain this by the gravity sliding hypothesis, but one would have to imagine the following series of movements:

1 The crust is depressed below sea level and on the ocean floor sediments are deposited.

2 First upward movement to the top of the crest envisaged in Haarmann's theory.

3 The sediments break free and slip down into the next depression, and fold upon themselves. The plane of slip is then inclined between 10 and 15° towards the depression.

4 The depression is uplifted, while the former crest goes

down, thus causing an inversion in the inclination of the slip-planes.

The oscillation theory therefore rests on rather a large number of unproven assumptions. We shall retain merely the fact that gravity can in some favourable situations play a part in facilitating the movement along thrust planes. Thus in the Ubaye region of the French Alps there are nappes which have moved further where facing one of the depressions than they have when facing the transverse crests of the mountain chain.

Currents in the mantle

We shall now look at another set of theories, those which ascribe the initiative to horizontal movement. When Wegener's theory of continental drift was first put forward it was immediately realised that the movement of the continents might provide a possible driving force for the folding of mountain belts. Folding due to such a cause might be expected to occur on the leading edges of drifting continents, of which the Andes and Rocky Mountains could be taken as examples, and in between two continental blocks which had moved together. An example of the latter might be the fold systems of the Mediterranean, the Atlas, Pyrenees, Alps and Apennine mountains, lying in the region between the African and European continents. Although, as we have seen, the mechanism of Wegener's original theory is not satisfactory, the idea that continental drift might be caused by convection currents in the mantle throws a different light on the problem altogether. If convection currents can carry the continents across the surface of the earth, then they may also provide the source of the lateral movements involved in folding.

The leading edge of a continent would not become buckled as the continent advanced if the continents are actually being carried by mantle material rather than moving through it, although the idea of folding due to two continents coming together may still be regarded as feasible. However, if there really are convection currents in the mantle it would be possible to have a fold belt at any place where two adjoining convection currents approach one another at the surface and turn down into the mantle, possibly at the margins of continents or in between them. In such a region, where the convection currents turn down from the surface, one would expect the crust, and the sediments lying on top of the crust, to crumple up as they were brought together on the opposing currents in the mantle. During the course of several million years a thick pile of folded sediments would accumulate as more and more crustal material and sediment was drawn towards the area of the downturn. Thus we have an explanation, not only of the folding, but also of the great thickness of sediments and the presence of a thick root of crustal material under regions of recent folding. A diagrammatic representation is shown in figures 59 and 62.

The formation of different fold belts in different regions of the earth during the course of geological time would necessitate occasional changes in the distribution of currents in the mantle. Should such a change take place in an area of folding so that it was no longer situated above a downward-moving current, then the downward drag on the sediments would cease, and isostatic forces would lift the upper levels of the fold belt above the surface to give an elevated mountain chain.

The convection current hypothesis explains a great many of the phenomena associated with the formation of fold belts, but there are other currents in the mantle which must also be

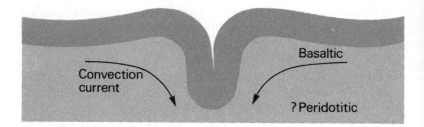

Figure 62. One view of the origin of fold belts. Convection currents in the mantle are thought to cause buckling and thickening of the overlying crust, which is predominantly granitic in composition.

considered. These are the currents which effect isostatic compensation. When the continents are eroded and the material removed from them is deposited in the oceans along their margins, there is a tendency for the continents to rise owing to their reduced weight and for the ocean floors to subside under their increased load. In such a case, equilibrium must be established by material returning from beneath the oceans to beneath the continents. It is of no great importance in the present context whether the movement actually takes place in the lowest crust or the upper mantle; what counts is the effect they have upon the crust. Either the currents will flow beneath the solid crust without affecting it in any way, or else there will be some deformation. The former alternative seems more likely in the case of rigid shield areas such as Scandinavia, which has risen en masse without folding since the melting of its ice cap about ten thousand years ago. If on the other hand, the crust should be more supple, containing areas of soft and easily deformed sediments, it is more likely that folding could take place as a result of the currents dragging against the lower part of the crust. If this supposition is correct it would provide an alternative explanation for the occurrence of fold belts along

Figure 63. The relation between transport of material from areas of erosion to areas of sedimentation and the compensating flow of the mantle material in the opposite direction.

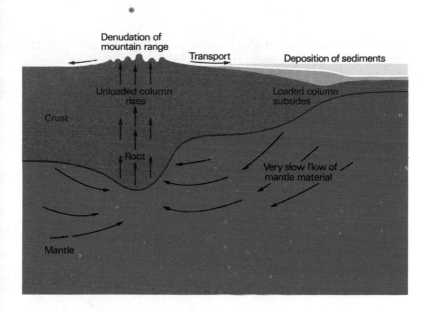

the edges of the continents, their arrangement in successive arcs, the crustal shortening, and even the apparent necessity for a minimum thickness of sediments: if they were not very thick the deep current would simply flow conformably underneath the crust without meeting any obstacle.

It was the American Dutton in 1889 who originated the hypothesis of folding through the currents of isostatic re-adjustment, but curiously enough it has hardly held any attention in America. There are several objections to which it is open. There has been no energetic or mechanical analysis of its workings and it seems to imply some sort of perpetual motion. It is difficult to see how a trench, once it has opened and filled with sediments can at one and the same time be a cause of a deep compensating current and the object of a compressive force induced by another similar current.

Having reviewed the theories which attempt to explain the folding of mountain ranges on earth we are now in a position to compare them. The theories of contraction and of vertical movement are ingenious, but none of them stands up to serious consideration. The most likely are those of convection and isostasy, both of which have their starting points in a state of non-equilibrium, of temperature in the case of convection, and of weight in that of isostasy. Since both kinds of current are known to exist, any final explanation must probably involve a combination of the two; perhaps the isostatic currents play some part in the localisation of the convection currents.

8 The future

There are many features of the anatomy of our planet which must remain open questions: the origin of the continents for example, or the causes of volcanic activity and mountain building. In answering these questions it will be necessary to consider the overall balance of things and the relative influence of external and internal forces – erosion and sedimentation on the one hand and deep seated currents and the transformation of crustal material at depth on the other.

Our knowledge increases greatly year by year, and will continue to do so. How can we be sure of this? Firstly, because the earth's physical properties are becoming better known all the time, mainly from the study of seismic waves, heat flow from the interior towards the surface, simple gravity measurements and more advanced interpretation of the results.

Laboratory experiments are providing us with new knowledge on the properties of rocks, their density, and characteristic changes of state. It is the production of high pressures which limits our advance here; leaks and possible accidents make these much more difficult to attain than the corresponding high temperatures which are also required.

We are being informed about our neighbouring planets and their properties by space craft. The density, temperature, shape, and the presence or absence of magnetic fields on Mars, Venus and the moon are all under investigation. We have just gathered some idea of how fruitful a comparison between these heavenly bodies and the earth can be, and these investigations will help us to understand our own planet just as much as the others. Artificial satellites are already feeding us with ideas about our own planet and in particular they have enabled us to measure its shape more accurately than was previously possible.

The interior of the earth, particularly enigmatic at the present time, will be a good deal better known when we are able to reach it, and extremely deep borings have been planned with this sole end in view. It is going to be a matter of passing through the crust and reaching if possible the mantle beneath.

The Russians reckon to site such a borehole on solid land, near to the sea and maybe on an island where seismic waves have shown the crust to be as thin as possible. It is thought from the preliminary reports of their distinguished scientist Beloussov that their shafts will penetrate to depths of between twelve and fifteen kilometres.

The Americans, on their side, have initiated the Mohole project, the 'Mo' part of the name being taken from the Mohorovičić discontinuity between the crust and mantle, which of course they must reach and penetrate. Their plan was to attack at the place where the crust is thinnest, the ocean floor under at least four thousand metres of water. Many of the problems involved have already been overcome. Their experimental drilling rig is mounted on the *Cuss 1*, a boat eighty metres long. The main difficulty is that of keeping the boat in one place on top of an opening four thousand metres beneath it, and a very satisfactory method has been developed. The boat is at the centre of a square with a buoy three hundred metres from it at each corner, firmly tethered to the ocean bottom by a cable with a two-ton block of concrete for ballast. Each buoy consists of an ultrasonic reflector, a radar reflector, a surface float, and another float one hundred metres under water, beyond the range and effect of the waves. The ship locates and maintains its position in relation to each of the buoys in turn, using both ultrasonic waves and radar, and performs any necess-

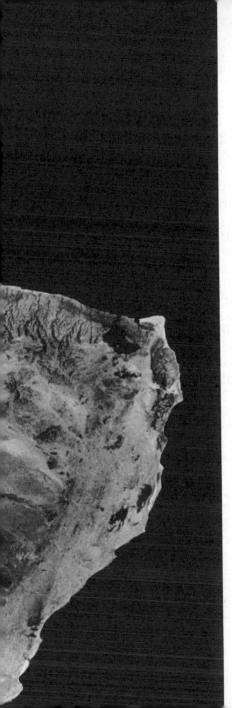

Figure 64. This picture of a view of the earth from space was taken from a spacecraft, and shows the south-eastern corner of Arabia. The sea appears black, the courses of the rivers (mostly dry) can be seen in the rugged mountainous area along the north-east coast, and in the south-west can be seen the long curving dunes of the sandy desert.

ary adjustments with four 200 h.p. outboard motors which are placed one pair forward and one pair aft. The drilling rig itself consists of an ordinary derrick thirty metres high, with the rods initially guided through a sort of bell-shaped cone suspended open end downwards to protect them from the ship's rolling and pitching, although an inclination of five or six degrees is admissible. A 1,700 horsepower motor drives the rods, and is built to take a 110 ton stress. The bit which actually makes the holes is a rotary diamond-studded variety, 22 centimetres in diameter.

Having tried out the experiment in fairly shallow water an experimental boring was carried out in the Pacific Ocean, 74 kilometres from the island of Guadalupe in water 3,566 metres deep. Sediments 183 metres thick were encountered first, and dated by their fossils as ten to twenty million years old. Then the drill penetrated into basalt and the crust of the deep ocean had been reached for the first time. The upper part of the basalt was vitrified and unmixed with sediment, which seems to indicate that it was extruded on to the ocean floor in direct contact with sea water, the first example of a submarine volcanic sheet in place. The heat flow and magnetic characteristics were also measured.

Although the trial boring was a great success, it has been decided not to carry on with the full scale attempt for the time being, the cost being too great. Nevertheless, in the face of such encouraging results we may hope that one day the project will be resumed. There are still a few problems to be solved. To reach the mantle there are five or six kilometres of basalt to be crossed, and a single drilling bit would not be sufficient; it would wear out and have to be withdrawn and replaced. How will they even find the hole again under four thousand metres of water? That problem is being considered.

The Soviet scientists are more fortunate in this respect because their apparatus will be more easy to manoeuvre.

Information on work of this kind is exchanged between the scientists of many different nations, and all the work is carried out under the wing of international scientific organisations and co-ordinated through them. There is a tremendous spirit of mutual good will, and thanks to this co-operation our knowledge of the earth is increasing faster than ever before. Let us hope that such cordial understanding will spread into other fields.

Ager, D. V., *Introducing Geology*, 243 pp., Faber & Faber, London 1961.

Bartels, Julius, ed., *Geophysik*, 373 pp., Fischer, Frankfurt-am-Main 1960.

Brinkmann, Roland, *Geologic Evolution of Europe*, translated edition, Enke, Stuttgart 1961.

Brinkmann, Roland, *Geologic Evolution of Europe*, translated from the German; condensed version of the second volume of the eighth edition of *Abriss der Geologie*, Enke, Stuttgart/Hafner, New York, 1960.

Cailleux, André, *Cours de Géologie*, Sedes, Paris 1962.

Cailleux, André, *La Géologie* (coll. 'Que sais-je?'), Presses Universitaires de France, Paris 1967.

Cailleux, André, *La Terre*, Bordas, Paris 1968.

Cailleux, André and Tricert, Jean, *Cours de Géomorphologie*, Sedes, Paris 1963

Carson, Rachel L., *The Sea around us*, 237 pp., revised edition, Oxford University Press, London and New York 1961.

Cayeux (=Cailleux), André, *Trois milliards d'années de vie*, 249 pp., Denoel, Paris 1964.

Clayton, Keith, *Earth's Crust*, 155 pp., Aldus Books, London/Doubleday, New York (for Natural History Press) 1966.

Cotton, C. A., *Geomorphology; an Introduction to the Study of Landforms*, 505 pp., seventh revised edition, Whitcombe & Tombs, London 1958.

Coulomb, Jean and Jobert, Georges, *The Physical Constitution of the Earth*, 328 pp., translated from the French, Oliver & Boyd, Edinburgh/Hafner, New York 1963.

Dauvillier, Alexandre, *Genèse, Nature et Évolution des Planètes*, 350 pp., Hermann, Paris 1947.

Dauvillier, Alexandre, *Cosmochimie et Chimie*, 215 pp., Presse Universitaire de France, Paris 1955.

Derruau, M., *Précis de Géomorphologie*, 413 pp., third edition, Masson, Paris 1962.

Dunbar, Carl O., *Historical Geology*, 500 pp., second edition,

Chapman & Hall, London/Wiley, New York 1960.

Dury, G. H., *The Face of the Earth*, 223 pp., Penguin, Harmondsworth 1959, Baltimore 1960.

Flint, Richard F., *Glacial and Pleistocene Geology*, 553 pp., Chapman & Hall, London/Wiley, New York 1957.

Fochler-Hauke, Gustav, ed., *Geographie*, 390 pp., second edition, Fischer, Frankfurt-am-Main 1962.

Fourmarier, P., *Principes de Géologie*, 1523 pp., third revised and enlarged edition, Masson, Paris 1950.

Furon, Raymond, *La Paléogéographie*, 405 pp., second completely revised edition, Payot, Paris 1959.

Gilluly, James and others, *Principles of Geology*, 534 pp., second edition, Baily Bros. & Swinfen, London/Freeman, San Francisco 1959.

Goguel, Jean, *Introduction à l'Étude Mécanique des Déformations de l'Éncorce Terrestre*, 530 pp., second edition, Imprimerie Nationale, Paris 1948.

Goguel, Jean, ed., *La Terre*, 1734 pp., Gallimard, Paris 1959.

Gutenberg, Beno, *Physics of the Earth's Interior*, 240 pp., Academic Press, New York 1959.

Haug, Émile, *Traité de Géologie. Vol. 1: Les Phénomènes Géologiques*, 538 pp., Armand Colin, Paris 1907.

Hill, M. N., ed., *The Sea. Vol. 3: The Earth beneath the Sea*, 963 pp., Interscience, London/New York 1963.

Holmes, Arthur, *Principles of Physical Geology*, 1288 pp., Nelson, London/Ronald Press, New York 1965.

Hoyle, Fred, *The Nature of the Universe*, 103 pp., new revised edition, Blackwell, Oxford/Harper, New York 1960.

Irving, E., *Palaeomagnetism and its Application to Geological and Geophysical Problems*, 399 pp., Wiley, New York 1964.

Jacobs, J. A., and others, *Physics and Geology*, 424 pp., McGraw-Hill, New York 1959.

Jaeger, Jean-Louis, *La Géochimie* (coll. 'Que sais-je?'), 120 pp., Presse Universitaire de France, Paris 1957.

Jeffreys, Harold, *The Earth*, 420 pp., fourth edition, Cambridge

University Press, Cambridge 1959.

Kettner, Radim, *Allgemeine Geologie. Vol. 1: Der Bau der Erdkruste*, 412 pp., translated from the Czech, Deutscher Verlag der Wissenschaften, Berlin 1958.

King, Lester C., *The Morphology of the Earth*, 699 pp., Oliver & Boyd, Edinburgh/Hafner, New York 1962.

Kirkaldy, J.F., *General Principles of Geology*, 327 pp., second revised edition, Hutchinson, London 1962.

Klebelsberg, R., *Handbuch der Gletscherkunde und Glazialgeologie*, Springer, Vienna 1949.

Kuenen, Ph.H., *Marine Geology*, 568 pp., Wiley, New York 1950/ Chapman & Hall, London 1951.

Lapparent, Albert de, *Leçons de Géographie Physique*, 718 pp., Masson, Paris 1898.

Leet, L.D. and Judson, S., *Physical Geology*, 502 pp., second edition, Prentice-Hall, Englewood Cliffs, N.J. 1958.

Levin, B.J., *Origin of the Earth and Planets*, 87 pp., translated from the Russian, second revised edition, Foreign Languages Publishing House, Moscow 1958.

Longwell, Chester R. and Flint, Richard F., *Introduction to Physical Geology*, 432 pp., second edition, Wiley, New York 1962.

Louis, Herbert, *Allgemeine Geomorphologie*, 354 pp., Walter de Gruyter, Berlin 1960.

Martonne, Émmanuel de, *Traité de Géographie Physique. Vol. 1: Notions gènérales. Climat. Hydrographie*, 496 pp., eighth revised and corrected edition, Armand Colin, Paris 1950.

Mason, Brian H., *Principles of Geochemistry*, 310 pp., second edition, Chapman & Hall, London/Wiley, New York 1958.

Mason, Brian H., *Meteorites*, 274 pp., Wiley, New York 1962.

Oparin, A. and Fesenkov, V., *The Universe*, 244 pp., translated from the Russian, Foreign Languages Publishing House, Moscow 1960.

Payne-Gaposchkin, Cecilia, *Introduction to Astronomy*, 508 pp., Prentice-Hall, Englewood Cliffs, N.J. 1954/University Paperbacks (Methuen), London 1961.

Read, H.H. and Watson J., *Introduction to Geology. Vol. 1: Principles*, 694 pp., Macmillan, London 1962/St Martin's Press, New York 1963.

Richter, Charles F., *Elementary Seismology*, 768 pp., Bailey Bros. & Swinfen, London/Freeman, San Francisco 1958.

Rittmann, Alfred, *Volcanoes and their Activity*, 305 pp., translated from the second German edition, Interscience, London/ New York 1962.

Romanovsky, V., and Cailleux, André, *La Glace et les Glaciers*, 128 pp., (coll. 'Que sais-je?'), Presse Universitaire de France, Paris 1961.

Runcorn, S.K., ed., *Continental Drift*, 338 pp., Academic Press, New York 1962.

Scheidegger, Adrian E., *Principles of Geodynamics*, 280 pp., Springer, Berlin 1958.

Schmidt, O., *Quatre Leçons sur la Théorie de l'Origine de la Terre*, 149 pp., Foreign Languages Publishing House, Moscow 1959.

Schwarzbach, M., *Climates of the Past*, 350 pp., Van Nostrand, New York 1963.

Shepard, Francis P., *The Earth beneath the Sea*, 275 pp., Johns Hopkins Press, Baltimore 1959/Oxford University Press, London 1960.

Sitter, L.U. de, *Structural Geology*, 552 pp., McGraw-Hill, London/New York 1959.

Smart, W.M., *The Origin of the Earth*, 224 pp., revised edition, Penguin, Harmondsworth 1959.

Strahler, Arthur N., *Physical Geography*, 534 pp., second edition, Chapman & Hall, London/Wiley, New York 1960.

Strahler, Arthur N., *The Earth Sciences*, 681 pp., Harper & Row, New York 1963.

Suess, Eduard, *The Face of the Earth*, 5 vols., translated from the German, Clarendon Press, Oxford 1904–24.

Supan, Alexander, *Grundzüge der physischen Erdkunde*, 852 pp., Von Veit, Leipzig 1903.

Termier, H. and G., *Histoire Géologique de la Biosphère*, 721 pp., Masson, Paris 1952.

Termier, H. and G., *L'Évolution de la Lithosphère. Vol. 2: Orogenese*, 940 pp., Masson, Paris 1956.

Thornbury, William D., *Principles of Geomorphology*, 618 pp., Wiley, New York 1954.

Trueman, Arthur E., *An Introduction to Geology*, 323 pp., second edition, prepared by J. A. G. Thomas, Murby, London 1964.

Umbgrove, J. H. F., *The Pulse of the Earth*, 358 pp., Nijhoff, The Hague 1947.

Urey, Harold C., *The Planets, their Origin and Development*, 246 pp., Yale University Press, New Haven 1952.

Vlerk, L. M. van der and Kuenen, P. H., *Logboek der Aarde*, second edition, W. de Haan, Zeist 1961. (French translation: *L'Histoire de la Terre*, Gcrard, Verviers 1961).

Wagner, Hermann, *Lehrbuch der Geographie*, 882 pp., Hahn, Hanover and Leipzig 1900.

Zumberge, James, *Elements of Geology*, 342 pp., second edition, Wiley, New York 1963.

Acknowledgments

Acknowledgment is due to the following for the photographs and diagrams (the numbers refer to the page on which the illustration appears).

11, 24–5, 64–5 USIS; 22 adapted from Griffith Taylor; 30, 59, 97, 114, 115, 130, 192–3, 224 Paul Popper Ltd; 30, 44, 82, 83, 213, 231 based on illustrations in A. Holmes; 41, 90 after illustrations in Jacobs, Russell and Wilson, *Physics and Geology*, McGraw-Hill, New York; 42, 106–7, 110, 120–1, 160–1 © Aerofilms Ltd; 48 (bottom) after K. B. Krauskopf; 50–1 after W. Q. Kennedy and J. E. Richey; 54–5, 62–3 © Solarfilma; 57 © Geological Survey and Museum; 66 after Dietrich; 68 C.O.I. Crown Copyright photo; 70 after B. Gutenberg and C. F. Richter; 75 after B. Galitzin; 81 US Geological Survey; 86–7 from map in Haroun Tazieff, *When the Earth Trembles*, Rupert Hart-Davis Ltd, London, 1964; 90 based on diagram in *Venture to the Arctic*, ed. R. A. Hamilton, Penguin Books Ltd; 109 courtesy Northern Ireland Tourist Board; 118–9 based on diagrams in Kirtley F. Mather, *The Earth Beneath Us*, Chanticleer Press, Inc NY, 1964; 122 and 123 based on illustrations in *Elements of Structural Geology*, E. S. Hills, Methuen, London, 1963; 134 and 135 after Sauramo 1939; 139 after H. H. Hess 1964; 202–3 based on David (1950), Hill (1958) and du Toit (1937); 206, 207 and 209 after E. Irving; 230 after Griggs; 237 NASA.

Index

World University Library

Some books published or in preparation

Philosophy and Religion

Christianity
W. O. Chadwick, *Cambridge*

Monasticism
David Knowles, *London*

Judaism
J. Soetendorp, *Amsterdam*

The Modern Papacy
K. O. von Aretin, *Göttingen*

Witchcraft
Lucy Mair, *London*

Sects
Bryan Wilson, *Oxford*

Language and Literature

The Birth of Western Languages
Philippe Wolff, *Toulouse*

A Model of Language
E. M. Uhlenbeck, *Leyden*

French Literature
Raymond Picard, *Paris*

Russian Writers and Society 1825–1904
Ronald Hingley, *Oxford*

Satire
Matthew Hodgart, *Sussex*

The Romantic Century
Robert Baldick, *Oxford*

The Arts

The Language of Modern Art
Ulf Linde, *Stockholm*

Architecture since 1945
Bruno Zevi, *Rome*

Twentieth Century Music
H. H. Stuckenschmidt, *Berlin*

Art Nouveau
S. Tschudi Madsen, *Oslo*

Academic Painting
Gerald Ackerman, *Stanford*

Palaeolithic Cave Art
P. J. Ucko and A. Rosenfeld, *London*

Primitive Art
Eike Haberland, *Mainz*

Romanesque Art
Carlos Cid Priego, *Madrid*

Expressionism
John Willett, *London*

Psychology and Human Biology

The Molecules of Life
Gisela Nass, *Munich*

The Variety of Man
J. P. Garlick, *London*

Eye and Brain
R. L. Gregory, *Cambridge*

The Ear and the Brain
E. C. Carterette, *U.C.L.A.*

The Biology of Work
O. G. Edholm, *London*

The Psychology of Attention
Anne Treisman, *Oxford*

Psychoses
H. J. Bochnik, *Hamburg*

Neuropsycho-pharmacology
A. M. Ernst, *Utrecht*

The Psychology of Fear and Stress
J. A. Gray, *Oxford*

The Tasks of Childhood
Phillipe Muller, *Neuchâtel*